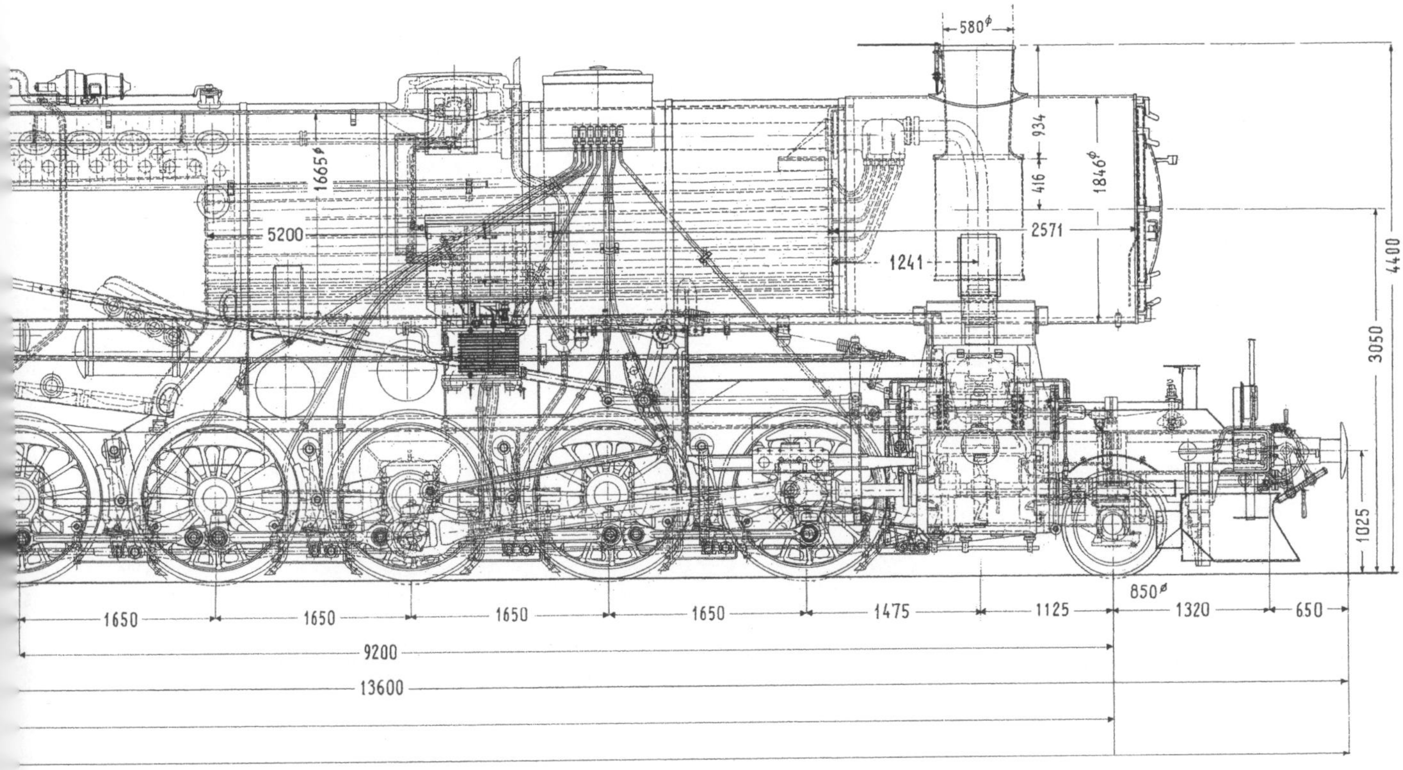
580ϕ
934
416
1846ϕ
2571
1241
1665ϕ
5200
4400
3050
1025
850ϕ
1650
1650
1650
1650
1475
1125
1320
650
9200
13600

Peter Slaughter, Alexander Vassiliev, Roland Beier

The German Class 52 "Kriegslok"

Frank Stenvalls Förlag

Preface

First of all, the authors thank Martin Murray, Volkmar Kubitzki, David Fakes and the late Sheridan Mazurek for providing essential information. Without their help this book would not have been published.

The authors would also like to thank the following people for all their contributions, which included provision of data, photographs and translations: Ilmar Adamson, Toms Altbergs, Zdenek Bauer, V. Belyaev, John Dawes, Dimiter Dejanow, P. Eilertsen, R. G. Farr, Roman Ficek, Witold Fita, Brian Garvin, A. Gogolin, Alfred Gottwaldt, Kurt Gozdal, Uwe Haas, Mehis Helme, Anna & Jack Hodgkinson, Reimar Holzinger, Wolfgang Kieslich, Lance King, Tomasz Kolodziejski, Allan de Koningh, Peter Kumelowski, Serban Lacriteanu, Tomas Meyer-Eppler, Bernd Neddermeyer, A. S. Nikolski, Jukka Nurminen, D. Trevor Rowe, John Scrace, Ian Simpson, Waldemar Sliwa, Mike Smith (Kelland photographic collection), Paul Spencer, Andreas Stange, Frank Stenvall, Brian Stephenson, Hamish Stevenson, Stephen Stoichkov, Chuck Sudetic, E. Talbot, Norbert Tempel, G. Villányi, Keith Wilde, Chris & Susanna Wilkinson, World Steam Magazine (David Thornhill), M. Wroblewski, two locomotive historians from Poland who preferred to remain anonymous, and anyone else we have inadvertently forgotten.

Finally thanks are offered to Helmut Griebl and Hansjürgen Wenzel whose book "Geschichte der deutschen Kriegslokomotiven" was the starting-point for this new publication.

Front cover: Russian Kriegslok TE-2505 on the Murmansk line in 1993 (photo David Fakes), PKP Ty2-998 at Sierpc 1977 (photo Sheridan Mazurek), DRB 52 3106 ca1943 (collection of B. Neddermeyer).

Page 1: Jaworzyna depot, Poland 1987. Photo P. J. S.

Published by Frank Stenvalls Förlag, Box 17111, S-20010 Malmö
ISBN 91-7266-140-2.

This book is simultaneously published in a German edition by the same publisher.

Printed by Skogs Boktryckeri AB, Trelleborg

Contents

Prototype locomotive 52 001 delivered by the Borsig works in September 1942. Photo courtesy Verkehrsmuseum Dresden.

Introduction

The German class 52 was one of the best-known and most widely used types of European steam locomotive. Over *six thousand* were built in a crash production programme from 1942 until the spring of 1945. In addition, several hundred more were completed after the war.

From 1945 onward, these "Kriegslokomotiven" (war locomotives) were taken over by most of the State railways of Europe, and became an everyday sight across much of the continent, from Belgium to Russia and from Norway to Turkey. They had been mass-produced with a minimum of skilled labour and scarce materials but proved to be a highly versatile design which gave many years service on all types of freight and local passenger work. At the end of the 1980s well over one thousand were still in existence, a few of them working in Poland and Turkey, whilst more than five hundred of the survivors were stored in more or less good condition in the Soviet Union and Bulgaria.

The epic story of the class 52 is generally well known, though at a detailed level many questions about them have remained unanswered. Much of the existing knowledge is described in the book "Geschichte der deutschen Kriegslokomotiven" written by Helmut Griebl and Hansjürgen Wenzel, published by Verlag Slezak in 1971. This lists all the class 52 locos which carried German numbers, and summarises the known history of every machine. When that book was written little was known about the locomotives of the SZD and PKP (Soviet and Polish state railways) so the post-1945 fate of over two thousand class 52s was still unclear.

In the last twenty years much additional data has come to light about the locos of the SZD, a lot of it copied from the Railway Museum at St. Petersburg which contains records of most withdrawn SZD locomotives. Details about the Kriegsloks which outlived the SZD into the 1990s have been collected by recent visitors to the surviving reserves of steam locomotives in the former Soviet Union, visits which would have been impossible for westerners until the end of the "Cold War". Meanwhile in other parts of Europe considerable progress has been made identifying the former German numbers of the Kriegsloks which worked on the Polish and Bulgarian State Railways.

It is the intention of this book to bring together all information now known about the class 52 and make it available to a wider readership. A large portion of the book is a table of all class 52 locomotives giving a very brief summary of their histories, with the purpose of showing where each loco has worked since 1945 and which locos remain unaccounted-for. Some readers are already familiar with the background story of the class 52, others are not, and for these an outline history is provided. The authors have not attempted to relate the entire "Kriegslok" story or answer all the outstanding questions as a more detailed book can be published in due course.

The class 52 has now virtually retired from everyday service, and there are already more than two hundred in the care of museums and preservation groups, or restored for hauling tourist trains. The time is now right

50 254 (Krauss-Maffei 15773/1939) illustrates the original class 50 design, soon developed into the class 50 ÜK and class 52. Krauss-Maffei works photo, from the collection of Alfred Gottwaldt.

to publish the information that enthusiasts have collected since Griebl and Wenzel published their book over twenty years ago. All the contributors hope this new book will bring as much pleasure to you in reading it as it has given us in producing it.

Origins

When the Deutsche Reichsbahn came into force in April 1920 it was one of the largest railway systems in the world. The majority of its extensive fleet of locomotives was inherited from the Royal Prussian Railways and included over five thousand locos of the modern freight classes $G8^1$, G10 and G12. Almost none of these engines was more than ten years old.

Production of new types of locomotives for the Reichsbahn started in the mid-1920s and until 1938 concentrated mainly on passenger and mixed traffic types. First of the Reichsbahn standard freight types was the 3-cylinder class 44 2-10-0 initially built in 1926 and mass-produced from 1937 onwards. These powerful locos had a 20-tonne axle-load limiting them predominantly to mainline use, and the same restrictions applied to the 2-cylinder variant (class 43) of which just 35 were built 1927-1928.

The second truly mass-produced type of Reichsbahn freight locomotive was the outstandingly successful class 50 2-cylinder 2-10-0, whose 15-tonne axle-loading and high maximum speed (80 kmph forwards and backwards) allowed widespread use over most of the Reichsbahn network on both passenger and freight services. The first class 50 was delivered in March 1939 and series production by Henschel, Krupp, Krauss-Maffei and WLF was in full progress before the end of the year.

After the outbreak of war in 1939 a shortage of freight locomotives developed which lead to a restriction of train services from January 1940 onwards. Steadily increasing numbers of class 50 locomotives were constructed during the years 1939-1941, together with smaller quantities of classes 03, 41, 44, 86 and other types. However, when the Reichsbahn had to provide supply trains for the invasion of the Soviet Union (which started 22nd June 1941) these were drawn from relatively outdated types of locomotives and rolling-stock. In the unusually severe winter of 1941/2 railway traffic to the eastern front was badly disrupted by drifting snow and and temperatures as low as -40°C. It is estimated that seventy percent of the German locomotives broke down, many damaged by the frost, and paralyzed trains were stranded far back into German territory. According to the Reichsminister for Transport (Dr.-Ing.e.h. Julius Dorpmüller) the Reichsbahn had so few operational locos and wagons that it could not meet the most urgent needs of transport even as late as May 1942.

Prototype locomotive 52 001 coupled to an early version of Wannen tender. Collection of Bernd Neddermeyer.

One of the many results of the transportation crisis was Hitler's demand for the construction of 15000 locomotives. Addressing the same problem on 7th January 1942, Göring (who was responsible for the current 4-year plan) decreed that locomotive production together with the mineral-oil industry should have priority over all other armaments projects. It would not have been possible to construct 15000 conventional engines at the required rate of 7500 per year, even using the available locomotive factories in the occupied territories, so after consideration of various austerity designs it was decided to standardise production on a radically simplified version of the class 50. Until factories could switch production to the new design (which became the class 52) most of the outstanding class 50 locomotives were completed with improvements for frost-protection but otherwise to progressively simplified standards. Austerity class 50s which were "Übergang zur Kriegslok" (transitional war locos) had the letters ÜK painted after their numbers from mid-1942 onwards. The first of these locomotives was 50 1583 which had been delivered by Henschel in March 1942 and lacked only smoke deflectors and warning-bell. Later class 50s were much simpler, so by summer 1943 the locos delivered by MBA were almost indistinguishable from the new class 52.

The prototype class 52 (52 001) was completed by the Borsig works on the 12th September 1942. Almost all the major locomotive manufacturers under German control were preparing to mass-produce this design, Henschel, Schwartzkopff and WLF leading the way with deliveries of more than 190 engines by the end of December 1942.

Design

The class 50 and the many variants of the 50ÜK and 52 form a family of locomotives with the same basic design and dimensions. To produce the plans for the class 52, the conventional Reichsbahn class 50 was redrawn to absolutely minimise the man-hours and skills used in its manufacture. Further design changes were adopted in response to wartime shortages of materials and to give the locomotives and their crews better protection against the Russian winter.

Design simplifications included

(1) removal of smoke-deflectors and front sections of running-plate

(2) removal of feedwater heater, feed-dome and one of the two sandboxes

(3) simplified application of firebox stays using progressively increasing amounts of welding

(4) replacement of Nicolai piston-valves with more straightforward piston-valves

(5) fabrication of connecting-rods, coupling-rods and other motion rods by welding

(6) simplified supports for valve-gear, with valve-spindle cross-head and slide replaced by a simple swing-link suspension

(7) use of disk-type pony wheels in place of the standard spoked wheel

(8) provision of austerity tenders, types K4T30 "Steifrahmen" and K2'2'T32 "Wannen".

Less successful simplifications included replacing the cylinder pressure-relief valves with breakable cast-iron disks, and omitting adjusting wedges for the horn blocks. In November 1943 the factories of Graffenstaden, Jung, MBA, Škoda, DWM and Krenau were ordered to re-introduce adjustable wedges to their production lines.

52 3119 delivered by Jung in April 1943. Typical of many early 52s, this loco had bar frames, additional frost-protection and a K4T30 rigid-frame tender. It is seen here at the factory where it was made, at Jungenthal near Kirchen an der Sieg. Collection of Bernd Neddermeyer.

It has been estimated that construction of each complete class 52 locomotive saved 6000 of the 17650 man-hours required to build a standard class 50. Manufacturers of the class 52 saved time in their own workshops and maximised deliveries by purchasing many components from outside suppliers, e.g. wheels, tenders and some of the boilers.

In order to save materials in short supply the bearing metal for axle-boxes, big-ends etc was changed radically from pre-war standards. The class 52 white-metal comprised 73.5% zinc and 10% tin, rather than the peacetime composition of 2% zinc and 80% tin. More conspicuously, most class 52 locomotives were constructed on 30mm thick plate frames, whereas the class 50s had bar frames made from 80mm thick steel.

Cold-weather protection measures included the totally-enclosed cab, and insulation for sensitive equipment such as the air-pump. Some of the earlier class 52 locomotives were provided with extra frost-protection and could be identified by their chimney-cowl and insulated tender (types K2'2'T26 and K4T30) in which the water could be warmed by steam-heating.

After the war, officials at the Esslingen and Krauss-Maffei factories told visiting engineers from Britain that although the central planning of the Kriegslok program had worked very well in the main, they were never able to "freeze" a design, and countless modifications had been issued to the works. Some of these had originated at other manufacturers and were designed to make production easier, whilst others came from the Reichsbahn and were to improve operating and maintenance.

Variants

Class 52 locomotives fall conveniently into two groups, those with standard plate-frame chassis and those with bar-frames as used on the class 50. Bar frames were eventually used on about one thousand 52s, predominantly those built before the middle of 1943. About the time production-lines switched to using plate-frames, other design changes were made including modifications to the suspension and mounting the air-brake pump on the boiler rather than on the chassis.

Turning attention to the tenders, most of the early 52s were paired with either K2'2'T26 tenders (similar to those used on the class 50s) or the K4T30 Steifrahmen variety. During 1943 the Wannen (bathtub) tender became standard because its simpler construction required only one third of the man-hours taken to build the class 50 tender, and moreover used less steel despite giving greater capacity for water and coal. Another change seen from summer 1943 was the adoption of lightweight smoke-deflectors, developed for the class 52 by Friedrich Witte and which used only one-fifth of the steel used for the class 50 deflectors. Early versions were made of straight plates with the top part angled inwards, but the familiar curved plates soon became standard. Small changes to the class 52 design were still being made late in 1944 and many of the last locomotives to be built (especially in France and Poland) showed simplifications to the cylinder casings and footplate supports.

The most important variant of the class 52 were the 178 condensing locomotives designed to cope with a lack of watering facilities in parts of the Soviet Union. In 1941 the Soviet troops in the course of their retreat had systematically wiped out most of the locomotive sheds, water-supplies and other technical equipment for the railways. The Henschel works had already had some experience of condensing locomotives as they had constructed a small number for Argentina as well as applying the principle to a 4-6-0 in Iraq and an E type 0-10-0 for the USSR. In the end, Henschel built all the condensing 52s and they proved both useful and reliable.

The design of the condensing locos allowed the exhaust steam to be recovered in the smokebox and condensed in a large 5-axle tender provided with cooling fans (tender type 3'2'T16). The locomotive's exhaust was driven by a fan, and this was in turn powered by steam leaving the cylinders and passing through a turbine. The system was self-regulating, as in a normal locomotive, and after the war it was applied to locomotives exported to South Africa where the distinctive whining noise made by the mechanisms is still remembered. The advantage of the condensing class 52 locomotives was their ability to travel up to 1000 km without taking water, but their cost was between 60 and 80 per cent more than standard engines. Particular operating difficulties were caused by the large 5-axle tender as it not only made the locomotive unsuitable for tender-first running but also too long to be turned on the large 23 metre turntables. A smaller 4-axle tender designated 2'2'T13.5 was used for deliveries from June 1944 (loco 52 1987) because the gradual retreat from the Soviet Union meant that shorter journeys were required.

Several experimental Kriegsloks were constructed to investigate quicker ways making boilers and fireboxes. For this purpose 50 3011 ÜK and 50 3012 ÜK were constructed with Brotan fireboxes, and 52 3620–52 3623 were provided with cylindrical stayless fireboxes. Rather more successful were seventeen locomotives 52 5058–52 5074 (and probably others) which had welded boilers. The all-welded boiler became widely used in several European countries during the next twenty years.

52 1852 at Berlin Schöneweide in March 1943, with train of newly-built condensing locomotives destined for the Eastern Front. Collection of Alfred Gottwaldt.

Manufacturers and construction totals

The number of class 52 locomotives completed before the end of the war in Europe is believed to be 6356. The greatest quantities were built by Henschel (1271) and the Wiener Lokomotivfabrik (1192), and the details for the other manufacturers are shown in table 1. Construction continued in small numbers after the war, and in the western zones of Germany 11 were made by Jung and 75 by Henschel including 9 condensing locomotives. In France Graffenstaden built a further 27 examples, while in Belgium where wartime plans to build 200 class 52s had come to nothing, 25 locomotives were made by each of the four manufacturers Cockerill, Franco-Belge, Haine St.Pierre and Tubize. The greatest number of postwar 52s were made in Poland where the Cegielski and Chrzanów factories completed probably 150. This figure, if correct, brings the grand total for the class 52 to 6719.

The final total of 6719 class 52 locomotives does not take account of locomotives on which construction was started but probably never finished, for example 52 7528 –52 7534 which were supposed to be under construction by Škoda in February 1945. Neither does the total include any locomotives rebuilt from war-wrecks, such as the 35 Soviet 8000 series locos from Ivano-Frankovsk works in the Ukraine.

It is worth mentioning that among the 28 (or more) subcontractors who made boilers for the class 52 there were some famous names, such as MAVAG who made 25 boilers in Budapest for Henschel, and Frichs who built ten at Aarhus in Denmark probably for Schwartzkopff. One notable name missing from the manufacturers in table 1 is Krupp whose factory at Essen received severe air-raid damage in March 1943 whilst

Table 1: Class 52 deliveries by manufacturer

	Wartime Construction 1942-1945					Postwar Construction 1945-1951						1942-1951
	DRB	CFR	HDZ	SDZ	TCDD	CFL	DB*	SNCB	SNCF	PKP	Ind	Total
Borsig	177											177
Cegielski	369 a									81 a		450 a
Chrzanow	372 a									69 a		441 a
Esslingen	281											281
Graffenstaden	139					10			17			166
Henschel	1197	40	9	15	10		73 d				2	1346
Jung	231						11					242
Krauss-Maffei	312 b											312
MBA	476	13										489
Schichau	592											592
Schwartzkopff	736	10	15									761
Škoda	153 e	17										170
WLF	1172	20										1192
Cockerill						c		25 c				25
Franco-Belge								25				25
Haine St.Pierre								25				25
Tubize						c		25 c				25
Total	6207 e	100	24	15	10	10 c	84	100 c	17	150 a	2	6719 a

Abbreviations:
Ind:Rheinische Braunkohlenwerke
* In the heading, the symbol DB covers not only the Deutsche Bundesbahn but also the railways of the American, British and French occupation zones (including the Saar territory) which were all incorporated into the DB between 1949 and 1957.
a: subject to confirmation.
b: This total is derived from the book "Geschichte der deutschen Kriegslokomotiven" by Griebl and Wenzel, page 196. However, the last three or four 52s from Krauss are sometimes counted as postwar locomotives.
c: Five of the locomotives from Cockerill and five from Tubize were handed over to the CFL when virtually new.
d: does not include 2 locomotives with Franco-Crosti boilers, 42 9000 and 42 9001.
e: including 20 locos operated by the BMB (railways of Bohemia and Moravia).
See also "6719 Kriegsloks Reihe 52", Eisenbahn (Wien) 4/1983 page 61.

Condensing locomotive 52 1863 in the Crimea in 1943. Collection of Alfred Gottwaldt.

they were still building the class 50ÜK. Although, as a consequence, Krupp were never able to complete any class 52 locomotives, they did supply class 52 boilers to other manufacturers. One example is boiler 3161/1943 (originally intended for a 50ÜK) which was carried by 52 5472 completed by Schichau in June 1943.

And now a few words about manufacturers' works plates. In keeping with the original austerity design, the first few thousand class 52 locomotives were delivered usually without any works plates, and instead had works numbers stamped into the chassis above the right-hand leading wheel. Most manufacturers later used very simple works plates welded onto the chassis in the same position. These works numbers have proved useful to railway-historians because from them it has been possible to work out the original German 52 XXXX identities of many locomotives seen in the early 1990s in Poland and Bulgaria.

First years in service

Class 52 locomotives were delivered to more than half of the Reichsbahn Direktionen (regions), predominantly to the central and more easterly areas such as RBD Halle, Berlin, Oppeln (Opole) and Königsberg, and a particularly large number also went to the RBD Wien. Over one thousand of the locomotives produced in 1942 and 1943 were allocated to the occupied eastern territories, where railways had a vital role as chief carrier of supplies for the war against the Soviet Union. Many of these locomotives were later brought back to Germany, though a growing number were captured by the Soviets and damaged by partisans. It should also be added that the railway network in Germany (and the occupied countries) became a priority target for air attacks in 1944, and one of the consequences was considerable damage to Reichsbahn locomotives.

149 class 52 locomotives were built for export in 1943 and 1944, and these were delivered to Rumania (100), Turkey (10), Serbia (15) and Croatia (24). Considerable numbers of 52s were also loaned to various countries, namely Turkey, Bulgaria, Rumania, Italy, Croatia and Hungary. Other 52s were loaned for use in occupied France, Belgium, Poland and the so-called Protectorate of Bohemia and Moravia.

The war in Europe came to an end officially on May 8th 1945, though the war in the Pacific continued for another four months. Within the new boundaries of Germany the railways were put under the administration of the four occupying powers, with coordination between the four occupation zones provided by a Transport Directorate forming part of the Allied Control Authority at Berlin. The Administrations and their Direktionen are as follow:

American zone (12982 km route-length):
Augsburg, Frankfurt am Main, Kassel, München, Nürnberg, Regensburg, Stuttgart.
Soviet zone (12675 km):
Berlin, Cottbus, Dresden, Erfurt, Halle, Greifswald, Magdeburg, Schwerin.
British zone (12354 km):
Köln, Essen, Hamburg, Hannover, Münster, Wuppertal.
French zone (7605 km):
Karlsruhe, Mainz, Saarbrücken.

The railways of the American and British zones were soon put under a unified management and received the name Deutsche Bundesbahn after the formation of the Federal Republic of Germany in 1949. The railways in the French zone were managed separately as the "Südwestdeutschen Eisenbahnen" from the summer of 1947 until 1952 when they became part of the DB.

Works plate from the frames of 52 4528.

Works plate from the frames of TE-293.

Works number stamped into the frames of 52 4853.

Deutsche Bundesbahn 52 138, delivered by Henschel in 1949 with three-drum feed-water heater. This was one of the DB's longest serving 52s, and lasted until May 1963. Collection of Bernd Neddermeyer.

Deutsche Bundesbahn 52 885 with Henschel feed-water heater, VTP-B 18000 Turbopump and extended chimney. Photo by Carl Spielmann.

West Germany

At the start of 1946 there were approximately nine hundred class 52 locomotives of wartime construction on the territory of the future Deutsche Bundesbahn. This total includes engines in the French zone and Saarland, and also includes those which had been damaged beyond economic repair [1]. Construction of 52s continued at Jung's works where eleven were produced 1945-1946, and at Henschel's works at Kassel where seventy-three were completed between 1945 and 1951. Among Henschel's postwar 52s, and not included in the above totals, were two which went to the Rheinische Braunkohlenwerke and two which were equipped with Franco- Crosti boilers and numbered on the Bundesbahn as 42 9000 and 42 9001 on account of their 17.6 tonne axle- loading. The quantity of 52s in West Germany grew still further between 1947 and 1951 with the return of over fifty locos from Italy, as well as three condensing locomotives from Belgium and one from France.

The 52s played a valuable role in the first years of post-war reconstruction, but in the longer term the Bundesbahn had little use for them as sufficient numbers of other freight classes were already available. Records for 1950 show that the locations with the greatest numbers of serviceable 52s were Friedrichshafen, Löhne, Mainz, Minden and Worms. To summarise the rapid disposal of the Bundesbahn's 52s, thirty-six were dispatched to Jugoslavia in 1951 / 1952, 41 examples built by Graffenstaden were returned to France, and the majority of the remaining 52s were officially withdrawn in 1954. In 1956 the Bundesbahn took out of traffic its last active standard 52 (52 2869 from Löhne depot) leaving in service only some of Henschel's postwar 52s which had been built with feedwater heaters.

In 1957 the Bundesbahn acquired fourteen more 52s from the Saarländischen Eisenbahnen [2]. This 552 km railway system based at Saarbrücken had been operated as an independent network from 1947, the same year that the Saar territory voted for economic union with France. Following a later plebiscite, the Saarland was incorporated into the Federal Republic of Germany with effect from 1/1/1957, and the railways were incorporated into the Bundesbahn on the same date. The fourteen class 52s based at Homburg depot are listed below, with a "z" against two locos which were already derelict.

52 1528, 52 2443, 52 2587, 52 3157, 52 3321, 52 3322, 52 3326, 52 3328, 52 3329, 52 3331, 52 5059, 52 5227(z), 52 5359(z), 52 6934

In 1963 the Bundesbahn withdrew its last more-or-less serviceable 52s, namely 52 138, 52 143 and 52 887. A small number were retained for heating duties including condenser locomotives 52 1895 and 52 1955 which survived at least until 1964. It is noteworthy that on the DB the class 52 was outlived by several older freight types, including the 57 (Prussian type G10) and class 50. The last of the DB's approximately five hundred 57s lasted until 1970, and the class 50, of which they originally had over two thousand, worked in diminishing numbers until the final year of steam traction, 1977.

It should be added that in the mid-1990s, long after 52s on the Bundesbahn had become a distant memory, visitors to the former Jugoslavian state of Bosnia discovered two former DB 52s in service near Tuzla, 33-236 (ex DB 52 817) and 33-248 (ex DB 52 4779). How many other former DB 52s continued to work in the 1990s is not yet known.

[1] The book "Geschichte der Deutschen Kriegslokomotiven" by Griebl and Wenzel shows 891 locos in the western zones of the Reichsbahn, not counting any loco-

Deutsche Bundesbahn 52 2025 leaves Minden with a freight train circa 1950. Completed by Henschel just after the war, this loco was paired with one of the 4-axle condensing tenders and was in service from October 1945 until 1953. Note the shape of the smoke plume from the fan-driven exhaust. From the collection of Alfred Gottwaldt.

motives delivered after the end of the war (8/5/1945) and not counting several Jugoslavian Kriegsloks which ended up in Germany.

[2] From 1952 the railways of this primarily coal mining area were called Eisenbahnen des Saarlandes.

Deutsche Bundesbahn 42 9000 near Köln Süd in June 1954. This loco was built by Henschel on the frames of the never-completed 52 893. It had a Franco-Crosti boiler with twin chimneys just in front of the firebox. Photo Kelland Collection, BRC.

DR 52 705 after "General repair", showing its feedwater heater and Giesl ejector chimney. The train is the 12:48 Personenzug P68562 Neubrandenburg to Friedland, 8th September 1969. Photo Brian Garvin.

East Germany

Archives suggest that at the start of 1946 there were at least 1325 class 52 locomotives on the Reichsbahn in the "Russian" zone of Germany. 485 are known to have been taken away 1946-1947 during the extensive programme of reparations to the USSR, most of these later working on the SZD (Soviet Railways) though a few went to work on Soviet industrial lines.

By 1949, after the most badly damaged locos had been officially withdrawn, there were just over eight hundred 52s left on the Reichsbahn, and the class remained the most numerous type of DR steam locomotive throughout the years of the German Democratic Republic 1949-1990 [1] [2].

The postwar DR kept the majority of its class 52 locomotives for about thirty years, but a small number were officially withdrawn 1946-1955 and thirty locomotives were sold to Bulgaria (20 in 1956, 10 in 1961). The DR acquired sixty 52s from the Soviet Union in 1962, including a number which had been taken as reparations in 1947.

Although most DR 52s remained without any fundamental alterations, rebuilding work by the Reichsbahn created two distinct sub-classes: the pulverised-fuel locos and the class 52^{80} "Rekoloks". The Reichsbahn's plans for converting some of its locos to burn pulverised

DR 52 3628 with rigid-frame tender type K4T30 at Leipzig Hauptbahnhof, 9th August 1967. Photo Brian Garvin.

brown coal were prompted by the DDR's plentiful supply of lignite and inadequate reserves of hard steam coal. Between 1953 and 1958 the Raw Stendal converted 25 class 52s to use the Wendler system for brown-coal firing. These locos were given 2'2'T24 Wannen tenders, modified cab and firebox, plus additional air-pump and reservoir to power the firing system [3]. The pulverised-fuel 52s worked from Senftenberg depot until finally taken out of service between 1974 and 1979, as shown in table 2. The last surviving example 52 4900 has been retained as a museum locomotive.

The class 52^{80} "Rekoloks" formed the second major sub-class on the Reichsbahn and comprised two hundred plate-framed class 52s reconstructed 1960-1967 with a new design of all-welded boiler (see table 3). The rebuilding took place in the latter part of the DR's extensive modernisation programme for its steam fleet, which had also included the rebuilding of 208 class 50s 1957-1962 and construction of 88 new class 50^{40} 2-10-0s between 1956 and 1960. The boilers of the 52^{80}s and the 50^{35}s (rebuilt class 50s) had the same design and featured a standard IfS/DR feedwater heater, shorter boiler tubes, a longer firebox and many other modifications.

Apart from the locomotives mentioned above, many DR 52s underwent wide-ranging and varied improvements whilst retaining the original design of the boiler. For example, probably at least 175 received a new rear-boiler unit built to the original design, but using welded construction, and more than 70 of these locos also received feedwater heaters. To describe many of these improvements the term "Generalreparaturloks" (GR) is widely used but without reference to a clear and precise definition.

It should also be mentioned that from about 1968 onwards the 52s were one of several classes to receive Giesl chimneys, though in general these lasted only until the middle or late 1970s.

In 1970 the Reichsbahn introduced a modified numbering scheme in which 93 class 52 locomotives received new identities. Pulverised fuel locos were renumbered into the 52 9xxx series so that, for example, 52 415 became 52 9415 and 52 4900 became 52 9900. Other locos between 52 006 and 52 872 were renumbered into the 52 1xxx series, as numbers starting with zero were reserved for oil-fired locos (and in fact all 52s remained coal-fired). Most renumberings comprised the prefixing of "1" to the loco number (eg 52 006 became 52 1006) but four exceptions were made to this rule to prevent duplication of numbers:

52 275 became 52 1175	52 436 became 52 1536
52 438 became 52 1538	52 626 became 52 1726.

Most of the un-rebuilt 52s survived well into the diesel era, with the geatest numbers withdrawn in the years 1975 - 1979. Fifteen were still on the DR's books in 1986 and happily the majority of these have now been preserved.

The 52s served on most Direktionen of the DR especially (in their later years) in the more central and south-easterly parts of the DDR. In 1980 the greatest numbers were at Berlin Schöneweide, Wustermark, Brandenburg, Röblingen, Frankfurt Oder, Elsterwerda, Falkenberg, Kamenz, Bautzen and Zittau. The class 52^{80} worked scheduled services until 1988 / 1989 when the DR virtually ceased regular steam traction on the standard-gauge. Thereafter they continued to serve as heating locos (Dampfspenders and Heizloks) and featured in a number of the "Plandampf" events. Other duties in the early 1990s included scheduled steam passenger services in the Berlin area, in particular the Nauen–Berlin–Lübbenau "Dampfgurke" which ran on summer Saturdays from 1991 until 1994.

[1] The authors know of 809 class 52s on the DR at the start of 1949. Of these, 23 were officially withdrawn 1949-1951.

[2] When counting the 52s as the most numerous type, the 50s and 50^{40}s have been considered to be separate classes.

[3] 52 415 had a K4T30 tender until 1967; 52 1580 and 1699 had tenders from G12 locos and until 1967 52 653 had a tender from a 17^{10} loco.

Table 2: DR Class 52^{90}

Original Number	Rebuilding Date	Number from 1970	Withdrawal Date
52 7195	07.08.53	52 9195	20.12.79
52 3222	09.12.53	52 9222	04.01.78
52 3250	08.06.57	52 9250	28.10.77
52 5278	26.11.57	52 9278	11.10.76
52 3285	24.03.53	52 9285	14.11.79
52 415	29.03.54	52 9415	01.07.77
52 1425	20.12.57	52 9425	14.11.75
52 1426	19.11.57	52 9426	29.07.76
52 3457	27.01.54	52 9457	17.03.76
52 533	03.01.58	52 9533	14.11.79
52 2543	16.07.57	52 9543	10.04.79
52 2576	05.08.57	52 9576	09.01.76
52 1580	07.07.53	52 9580	25.10.76
52 3594	24.03.53	52 9594	24.03.77
52 2650	23.07.53	52 9650	25.10.74
52 653	15.07.53	52 9653	06.03.79
52 2681	57	52 9681	23.04.75
52 1699	16.10.57	52 9699	06.03.79
52 716	24.01.58	52 9716	14.05.79
52 2718	31.12.57	52 9718	02.01.80
52 1737	12.12.57	52 9737	20.04.77
52 5762	01.04.54	52 9762	22.12.78
52 3794	31.12.53	52 9794	19.05.76
52 2818	05.09.57	52 9818	10.04.79
52 4900	30.04.54	52 9900	Preserved

DR 52 415 with pulverised-coal firing at Cottbus, 10th September 1969. Photo Brian Garvin.

DR 52 5918 with a type 2'2'T26 tender at Dresden depot on 25th July 1973. Photo Hamish Stevenson.

DR Rekolok 52 8005 on an eastbound freight train at Arnsdorf, 26th August 1982. This loco has tapered sandboxes, and note also the edging around the smoke-deflectors which is a little unusual for a class 52^{80} locomotive. Photo P.J.S.

Table 3: DR Class 52^{80}

52 8001–8200 were rebuilt from the class 52 locomotives shown below between 1960 and 1967.

	1	2	3	4	5	6	7	8	9	10
52 8001-8010	671	2654	6357	2616	1111	2644	2560	7605	1605	7172
8011-8020	1569	4944	3734	5207	3464	2886	3916	1572	1711	627
8021-8030	6684	6336	2706	2542	1421	3257	7247	563	5018	5012
8031-8040	3550	1252	539	7583	5761	1617	712	5274	2720	1152
8041-8050	5243	7122	358	3455	5789	718	6359	1499	3721	704
8051-8060	6779	5121	2551	3633	1649	6778	7248	7717	4876	3193
8061-8070	5118	2467	1108	6011	6415	6372	7003	3311	2199	3223
8071-8080	1402	2387	7771	6427	1292	375	1454	6996	5659	5015
8081-8090	373	5190	3699	1710	4512	2295	2455	3169	2649	7778
8091-8100	7223	1604	711	4501	6106	2312	2701	3420	1443	1145
52 8101-8110	5260	7550	3673	5584	614	6159	6816	7109	2883	3403
8111-8120	5992	3318	1159	819	3568	3354	7221	2180	1282	2652
8121-8130	5585	6390	1633	2501	6300	362	2177	3473	1325	1143
8131-8140	3218	7789	7793	7138	474	1175	5803	7313	3209	470
8141-8150	5315	4907	2289	1648	1169	6873	2648	547	3839	7242
8151-8160	2696	6082	7284	4896	1714	5678	6829	346	3166	532
8161-8170	5519	7613	5996	6277	5250	578	7251	3574	1248	3254
8171-8180	1514	1603	7734	2453	3211	3871	4996	1501	2763	3625
8181-8190	1254	472	2656	3722	2858	632	2626	3590	5306	2887
8191-8200	2675	493	6339	556	4901	5374	7581	1217	3252	467

Years of reconstruction:

1960: 52 8001-8008
1961: 52 8009-8038
1962: 52 8039-8071
1963: 52 8072-8101
1964: 52 8102-8122
1965: 52 8123-8154
1966: 52 8155-8180
1967: 52 8181-8200

Top: *52 8200 at Löbau with a westbound freight train, 28th August 1982. Photo P.J.S.* ***Centre:*** *DR Generalreparaturlok 52 3548 leaves Cottbus with a freight train for Frankfurt/Oder in October 1984. The loco is running tender-first, but this is not clear from the photograph because a strong wind is blowing from the west, and the train is being overtaken by its own smoke. Photo P.J.S.* ***Bottom:*** *DR 52 1006 on a local passenger train in Dresden, 27th July 1972. Delivered as 52 006 in October 1942, this was the DR's lowest numbered 52. It carried the number 52 1006-7 from 1970 until withdrawn from service in September 1977. Photo Hamish Stevenson.*

DR 52 8171 with Giesl ejector and 52 1715 at Stendal depot in March 1972. Photo Brian Stephenson.

DR 52 1146 thunders through Dresden with a freight train, 28th July 1972. Photo: Hamish Stevenson.

Norway

German locomotives taken to Norway during the war included the class 52, 74 of which later became class 63a on the Norwegian State Railways (NSB). Some of these were withdrawn in the 1950s but the majority survived to 1963 - 1964, and the last examples were not withdrawn until 1970. They are perhaps best remembered in the Trondheim area and on the line from Hell to Mo i Rana.

In 1972 four members of the class were recovered from a strategic reserve in Drangsdal tunnel where they had been going rusty since 1958. Of these survivors two were acquired by museums, 2770 going to Hamar and 5865 going to England for display at Bressingham.

The class 63a retained their original German numbers and comprised the following locomotives:
217, 252, 324, 660, 1100, 1101, 1104, 1106, 1107, 2293, 2343, 2570, 2572, 2724, 2770, 2843, 2863, 3113, 3437, 3606, 3758, 3832, 4755, 4830, 4831, 4832, 4833, 4834, 4835, 4836, 4837, 4838, 4839, 4929, 5032, 5081, 5113, 5116, 5133, 5321, 5371, 5397, 5573, 5606, 5664, 5810, 5839, 5841, 5842, 5843, 5844, 5845, 5846, 5847, 5848, 5849, 5850, 5851, 5852, 5853, 5854, 5855, 5856, 5857, 5858, 5860, 5862, 5863, 5865, 6081, 6204, 6266, 6396, 6412

Top: Former 52 4834 heads a freight train on the Norwegian State Railways, probably in the mid 1950s. Photo W. J. Verden Anderson, courtesy of Rail Archive Stephenson.

Norwegian State Railways 5664 at Krossen in May 1962. Photo Kelland Collection, Bournemouth Railway Club.

SNCB 26.048 at Latour depot in 1953. Photo Dr P. Ransome-Wallis, National Railway Museum, York.

Belgium

During the occupation of Belgium there were plans to build two hundred German class 50ÜK locomotives and two hundred class 52s. 186 of the 50ÜKs were duly delivered by August 1944, and the remaining fourteen were completed after the liberation ultimately becoming class 25 on the Belgian National Railways (SNCB). One hundred of the "class 52" locomotives were completed by Belgian manufacturers in the early postwar years and the type became SNCB class 26. (The remaining hundred were never built). In 1946 ten of the class 26s were hired to CFL in Luxembourg who purchased them not long afterwards. Details are summarised in the table below.

Class 26 locomotives were characterised by shortened chimneys, bar frames and a design of open-back cab not seen on any other 52s. There were also various minor differences between the locos from the four manufacturers. Two types of tenders were used corresponding to the German designs 2'2'T26 and K2'2'T32 (Wannen), and both types were provided with a weatherboard for tender-first running. Three class 52 condensing locomotives were left behind in Belgium by the German forces. These were 52 1973, 52 1977, 52 1992 which were allocated SNCB numbers 27.003, 27.001 and 27.002 before returning to Germany in 1950.

The 26s worked from depots Bertrix, Kinkempois (Liège), Montzen, Latour, Schaerbeek (Brussels) and Renory. The first fifty locos to be withdrawn went at the same time as the class 25s (1955 - 1959), while the remainder survived in dimishing numbers until 1963.

At the time of writing there are four locomotives preserved in Belgium which are related to the classes 25 and 26:

50 3666	ex DR, rebuilt from 50 2145, works number Franco-Belge 2567 / 1944.
"26.101"	ex PMPPW (Poland). Originally 52 3554, built by Krauss Maffei in 1943.
52 3314	ex ÖBB, built by Jung in November 1944.
52 8200	ex DR, rebuilt from 52 467, originally completed by Borsig in 1943.

Builder	Works Number	SNCB Nr. to 1946	SNCB Nr. from 1946	CFL
Tubize	2308-2319	2600-2611	26.100, 26.001-26.011	
	2320-2324	2612-2616	26.012-26.016	5601-5605
	2325-2332	2617-2624	26.017-26.024	
Cockerill	3338-3354	2625-2641	26.025-26.041	
	3355-3359	2642-2646	26.042-26.046	5606-5610
	3360-3362	2647-2649	26.047-26.049	
Haine St.Pierre	1914-1938	2650-2674	26.050-26.074	
Franco-Belge	2623-2647	2675-2699	26.075-26.099	

SNCB 26.059 plus 25.003 (on the left) and 26.093 plus 26.094 double-head iron-ore trains at St Mard in 1953. Photo Dr P. Ransome-Wallis, National Railway Museum, York.

Luxembourg

During 1946 - 1947 the state railways of Luxembourg (CFL) took delivery of twenty class 56 locomotives (German class 52 design), of which ten were purchased more or less new from Belgium and ten came direct from the Graffenstaden factory in France. These were not the only new freight locomotives delivered after the war, as twenty class 55s (German type 42) were delivered from WLF and ten class 47 "Liberation" 2-8-0s came from the Vulcan Foundry in Britain.

The class 56 locos were numbered as follows:

5601-5605	Tubize	2320-2324/1946
5606-5610	Cockerill	3355-3359/1946
5611-5620	Graffenstaden	8044-8053/1946

The Belgian locomotives were originally constructed as SNCB 2612 - 2616 and 2642 - 2646 and they arrived in Luxembourg carrying their Belgian numbers. All had bar frames and open-backed cabs with cab-ventilators reminiscent of the German class 50. On the other hand, the Graffenstaden machines were more typical of the standard "Kriegslok" design, having plate frames and enclosed cab, though a new peacetime luxury was the provision of an extra side window on each side of the cab. In order to improve the efficiency of class 56 locomotives the CFL equipped them with ACFI feed-water heaters housed in two long drums on each side of the chimney.

Initially the 56s were stationed at Luxembourg and Trois Vierges depots, working the main line from Luxembourg city northward toward Gouvy in Belgium. On this route they operated many of the passenger services including international expresses, as well as heavy coke and coal trains. General Motors Co-Co diesels (CFL type 1600) took over the express workings in 1955 plus some of the freight traffic, and further passenger services were dieselised in 1957.

Other workings by the 56s were from Luxembourg city eastward to Wasserbillig and westward to Arlon in

CFL 5609 in October 1956. This loco was built by Cockerill in Belgium in 1946. Part of the ACFI feedwater heaters can be seen next to the chimney, and the feedwater pumps can be seen above the cylinders. Kelland Collection, Bournemouth Railway Club.

Belgium, and (from 1958) various duties from Pétange depot where they replaced the class 54 (Prussian type G12).

All 56s were withdrawn in 1961, though four were kept in reserve for engineering trains until 1964 which was effectively the last year of steam traction on the CFL.

France

The original construction programme for the class 52 specified that 250 examples were to be built at Graffenstaden, near Strasbourg, at the works of Société Alsacienne de Constructions Mécaniques (SACM). At the time the 52s were built Alsace had been incorporated into Germany and the factory is often referred to as the Lokomotivfabrik Graffenstaden. A total of 139 52s were completed here between June 1943 and October 1944, followed after the war by seventeen locos for the SNCF and ten for Luxembourg. The works numbers of these locos are shown in Appendix A.

On the SNCF the German class 52 became the class 150Y and numbers 150Y1 to 150Y17 were assigned to the postwar SACM locos, though at first they carried their intended German numbers 1739-1755. The SNCF also took over twenty-five former German 52s which it operated under their Reichsbahn numbers prefixed 150Y. The locos were:

150Y123, 244, 1306, 1593, 1637, 1657, 1673, 1677, 1678, 1993, 3125, 3149, 3483, 3536, 3595, 3697, 3811, 5687, 6178, 6848, 6982, 7283, 7444, 7601, 7604.

A noticeable feature of the 150Ys was the chimney which was 20cm shorter than the standard design in order to fit the French loading-gauge. Postwar SACM locos had distinctive flat-topped cylinder-casings, a design also seen on locos delivered by Franco-Belge and on Polish 52s built from late 1944 onward.

Forty-one other 52s which had been built by Graffenstaden were returned by West Germany a few years after the war but these were not put into service. The SNCF clearly had an adequate number of mixed-traffic locos at this time, notably 1323 of the 141R class newly delivered from the USA and Canada.

The 150Ys worked in Alsace and Lorraine on the Eastern region of the SNCF. They were allocated to depots Mulhouse Nord, Mulhouse Île-Napoléon, Sarrebourg, Sarreguemines, Thionville and Hausbergen (Strasbourg). As in so many other parts of Europe their duties included both heavy freight trains and passenger services. On retirement from the SNCF, four locomotives went to work in the coal industry for the Houillères du Bassin du Nord et du Pas de Calais (HBNCP). The four locos were 150Y-3536, 150Y-16, 5 and 9 – subsequently numbered by their new owners as Y1–Y4.

The 150Ys could be seen on the SNCF until 1957 and those in the coal industry lasted until 1961.

Italy

Over one hundred class 52 locomotives are known to have been used by the German forces in Italy in 1944 and 1945. After the end of the war some of these locos continued to work on the Italian State Railways (FS) in the Verona and Udine areas. They were returned to Germany between 1944 and 1951.

In the summer of 1945 the British authorities in northern Italy borrowed ten Austrian 52s and used them for hauling supply trains going to Austria on the line from Venice to Udine. The 52s were handed over to the FS for operation and were returned to Austria in 1947.

In more recent years another 52 has found its way to northern Italy. This is Jugoslavian state railways 33-107 (ex 52 4752) which is now on show at the railway museum at Trieste.

CFL 5618 in September 1959. This loco was built by SACM in France in 1946. Note the flat-top casings of the cylinders. Photo Kelland Collection, Bournemouth Railway Club.

ÖBB 52.6766 and 52.6964 depart from Hieflau with a heavy train for Selzthal, 24th June 1966. Photo Brian Stephenson.

Austria

During the Second world war Austria was not a sovereign state but part of the German Reich, and the state railways of Austria were incorporated into the Reichsbahn. The Wiener Lokomotivfabrik (WLF) in Floridsdorf took part in the manufacture of the class 52, like almost all the other loco works under German control, and deliveries finally totalled 1192 locomotives. From the WLF also came the design for the K4T30 rigid-frame tender. A considerable number of K4T30s were constructed at WLF's Raxwerk factory in Wiener Neustadt before production was switched to the standard "Wannen" tenders.

Class 52 locomotives were already being used in Austria in the spring of 1943, primarily in the Reichsbahn "Direktion" of Vienna. Smaller numbers were also allocated to Direktions Villach and Linz. In 1944 the class 52 was also stationed in Wörgl and Innsbruck, on account of the bomb damage to the electrified lines in those areas.

In April 1945 the Austrian government was re-established and the Austrian railways were constituted as the Österreichische Staatseisenbahnen (ÖStB). At this time the number of 52s in Austria totalled about 700, about half of which were in the east of Austria, in the Soviet occupation zone. The ÖStB became the Österreichische Bundesbahnen (ÖBB) in 1947.

On the instructions of the British authorities, ten 52s were loaned to Italy from August 1945 until 1947. In addition 32 locos were loaned to the West German "Bizone" in 1947 and 1948. Of the stock outside the Soviet occupation zone of Austria, 28 locomotives had been written-off before 1953.

In the Soviet zone the 52s were considered as booty locomotives, and were given a T painted in front of their numbers. (T = Trofeinyi = Captured). Between April and July 1945 probably more than 150 were taken away by the Soviets, many going to Hungary, and in the following years smaller numbers were handed over to Rumania (1946) and Jugoslavia (1947). Most of the remaining 52s were taken away to the USSR in December 1948. Only eleven machines remained behind, these being the

ÖBB 52.2458 with oilfiring at Wien Ost, 27th June 1956. Photo Kelland Collection.

ÖBB 52.859 with Giesl ejector chimney at Wien Ost in October 1972. This locomotive was made by Krenau in October 1944, and like other 52s made in Poland at that time it has flat-top cylinder casings. Photo Kurt Gozdal.

ÖBB 52.3319 with Giesl ejector and cabin-tender on a freight train at Oberlaa, 22nd September 1972. Photo Kurt Gozdal.

so-called "Sperrlokomotiven" at Wien Ost. These went to the East in March 1953.

Remaining on the ÖBB were just over 300 class 52 locos, of which 36 had bar frames (reclassified as ÖBB class 152 in 1953) and 271 had plate frames (ÖBB class 52).

Owing to the shortage of coal in Austria in the immediate post-war years , the ÖBB operated some of its locos with oil-firing until the mid 1950s. Later all 152s and about 100 52s were given Giesl chimneys, and about 40 locos were also given Heinl feedwater heaters. By these means the efficiency of the locos was increased and greater economy was achieved. Finally it should be added that a few ÖBB locos were experimentally equipped with Trofimov piston valves.

Further rebuilding work related to the tender. Numerous Wannen tenders were given guards' compartments which made it unnecessary to add a separate guards' van for freight trains. This feature had originally been tried-out on a prototype built by Westwaggon in 1942.

The 52s and 152s formed the back-bone of the ÖBB's locomotive fleet until procurement of diesel locomotives and progressive electrification lead to their gradual displacement and withdrawal.

After 1970 they were still in service from Linz towards the Gesäuse (St Valentin–Selzthal) and on the Pyhrnbahn (Linz–Klaus–Selzthal) and also on the line to Summerau. They were also still used on the non-electrified lines in and around Vienna. The last "reservation" of the class 52 was the Weinviertel area north-east from Vienna. The depots of Mistelbach and in the end Straßhof used their 52s primarily for local services.

In August 1976 there was a strong increase in ÖBB goods traffic due to the collapse of the Reichsbrücke in Vienna (a major road bridge across the Danube). The ÖBB had to consequently reactivate some 52s and use them on the Nordbahn (Vienna–Hohenau–Breclav). The final end for the 52s was on the 31st December 1976, and on this date the ÖBB ended its standard-gauge steam traction – with the exception of rack locomotives working on the Erzbergbahn.

Some 52s remained for a while as a strategic reserve, but ultimately these went the way of all old iron except a number which were preserved. It should be mentioned that the frames of 35 locos were rebuilt for use as Klima snow-ploughs and are still in service with the ÖBB (class 9760, formerly class 985.2).

Between 1968 and 1973 thirteen 152s were sold to the Graz–Köflacher-Bahn (GKB), a railway in western Styria. On the GKB they replaced the considerably older class 56 locos and were used both on freight traffic and also peak-hour passenger trains. Dieselisation however made the 152s redundant here too, and they were last used in April 1978.

Another railway which used the class 52 was the Raab–Ödenburg–Ebenfurther Railway (GySEV). This Austro-Hungarian organisation centred on Sopron (Ödenburg) acquired six class 520s in 1963/4 from the Hungarian State Railways (MAV), which had in turn received them from the USSR in 1962/3. In 1976 two more 520s followed to Sopron. At first the locos had the large Soviet headlamp (later replaced by smaller lamps) and they were also characterised by the small Soviet smokebox door.

For a while the 520s operated all the freight traffic between Sopron and Ebenfurth. The gradients on this line often demanded the use of banking and double-heading. Scheduled use of steam traction ended in February 1979 when the 520s were replaced by M62 diesels, though for a while 520s still saw some use when diesel locomotives failed. In October 1981 a 520 headed the local goods train from Sopron to Ebenfurth, and this was the last time a 52 was used in normal service in Austria.

Klima snowplough 9760 019 at Wien Ost depot, 10th June 1995. This was built on the frames of 52 6969 in 1970. Photo Roland Beier.

GySEV 520,094 at Neufeld a. d. Leitha, 5th July 1970. Photo Frank Stenvall.

Hungary

During the second world war production of new locomotives by MAVAG in Hungary concentrated on the class 424, with a total of more than two hundred completed between 1940 and 1944. MAVAG also had an order from Henschel to build twenty-five class 52 boilers, and these were duly completed from March to December 1944. To meet the high levels of wartime traffic, Hungary hired large numbers of locomotives including German class 52s. Serviceable and repairable 52s still in Hungary at the end of the war were taken away by the Soviet authorities between 1945 and 1950.

In the 1950s the principal mainline freight locomotives on the Hungarian State Railways (MAV) were the approximately 500-strong class 411 which were former American S160 war locomotives. These were supplemented circa 1958 by the purchase of five second-hand Austrian 42s and by using a few ex-German 52s hired from the USSR. The latter locomotives (listed in table 4b) were broad-gauge and used at Záhony on the border with the Soviet Union.

In 1963 one hundred former German 52s were acquired from the USSR becoming MAV class 520. Ninety-four were converted to standard gauge and received numbers 520,001–520,094; the remaining six were used on the broad-gauge at Záhony and were numbered 520,5001–520,5006. The 520s saw quite widespread use on the MAV's freight traffic although they were considerably outnumbered by the 411s and the mixed-traffic 424s. By 1970 many of the heavier freight trains were powered by Soviet-built M62 diesels, but a few MAV

Table 4a. MAV Class 520 and their original identities

	1	2	3	4	5	6	7	8	9	10
520,001-010	52 441	2287	3177	5219	6006	7072	7335	710	2390	3310
520,011-020	52 4987	5598	7311	1799	2315	2499	4918	7443	7558	3897
520,021-030	52 4802	5576	6302	6788	7505	1526	1622	1684	2805	3535
520,031-040	52 5022	5307	6167	6253	6386	7357	7658	1249	2302	2357
520,041-050	52 2625	2716	5069	5607	7331	7559	7761	2217	3443	3815
520,051-060	52 3823	5127	5156	7149	7786	672	776	1178	1625	2170
520,061-070	52 2189	2198	2223	2331	2466	2834	2884	3610	3767	5556
520,071-080	52 6128	7564	7754	7772	464	2239	2573	3528	6944	3735
520,081-090	52 4515	5128	5595	7686	870	2451	3258	3762	4411	4547
520,091-094	52 4758	6002	7160	7781						
520,5001-5006	52 286	665	7777	4535	5392	7791				

Note: It is said that seven withdrawn 520s were reinstated in 1972 to replace locomotives in worse physical condition. The reinstated 520s were given the numbers of the locos being replaced, as shown below:

520,005" = 520,057	520,006" = 520,071
520,023" = 520,049	520,028" = 520,007
520,034" = 520,053	520,036" = 520,086
520,065" = 520,069	

520s could still be seen in certain areas such as the Budapest to Szob line, and at Celldömölk and Komárom. Among the last survivors was 520,075 which shunted in the marshalling yards at Hatvan until the mid 1980s.

Four 520s were taken over by the Györ Sopron Ebenfurti railway (GySEV) in 1963, and four more were taken over up to 1976 as detailed in table 4c. They were used on the GySEV's freight traffic between Sopron and Ebenfurth, and were supplemented by MAV 520s for autumn sugar-beet traffic. The 520s on the GySEV were replaced by M62 diesels and were officially withdrawn between 1979 and 1984.

After the effective end of MAV steam traction in 1985 a number of 424s and 520s were retained as "stationary boilers" and renumbered accordingly. 520,067 became number 520-01 (located at Debrecin in 1992) and 520,074 at Budapest Hámán Kató shed was renumbered as 520-02. Three 520s have been saved for preservation, namely 520,030 in Fertöboz, 520,034" in Komárom and 520,075 in Hatvan.

Table 4b: Broad-gauge Soviet locomotives on loan to Hungary 1957-1958 for use at Zahony:

52 208, 870, 4520, 4853, 4942, 5312, 5638, 5895, 8008, 8028, 9102

Table 4c: GySEV Class 520

520,018	5.1964-1.10.1979	+ 1.10.1979
520,020	5.1964-1.10.1979	+ 1.10.1979
520,030	1.1.1976-13.9.1984	preserved
520,050	9.10.1963-26.3.1980	*
520,075	1.1.1976-15.11.1979	to MÁV
520,079	12.9.1963-15.6.1980	+ 15.6.1980
520,083	12.9.1963-16.8.1981	+ 16.8.1981
520,094	29.10.1963-1.10.1979	+ 1.10.1979

* sold as boiler to Zalaszentgrot.

Table 4d: MAV 520s loaned to GySEV for autumn sugar-beet traffic.

520,003	1970
520,025	1969
520,027	1971, 1977
520,029	1970, 1977
520,030	1968, 1969,
520,039	1969
520,044	1963, 1965
520,046	1965
520,048	1964, 1969
520,049	1968, 1969, 1970
520,063	1965
520,072	1970
520,092	1969, 1970
520,093	1969

Hungary: MÁV 520,075 - originally 52 464. Photo Sheridan Mazurek. ***Below:*** *Former Soviet TE-7558 (later renumbered MÁV 520,019) on a freight near Cegléd in May 1963. Photo D. Trevor Rowe.*

CSD 555.3054 leaves Bratislava for Breclav, 15th August 1967. Photo Brian Garvin.

Czechoslovakia

During the years of occupation (1939-1945) Czechoslovakia was divided into the "Protectorate of Bohemia and Moravia" and a separate state of Slovakia, plus extensive territories which were incorporated into the German Reich and Hungary. The Skoda factory at Pilsen built a large number of German locomotives in this period, delivering 140 of the class 50 followed by 153 of the class 52. In addition Skoda built 17 locos of the class 52 design for export to Rumania.

The railways of Bohemia-Moravia supplemented their own locomotives by leasing engines from Germany. These included 52 7475–52 7494 which were delivered by Skoda from May to July 1944. The railways of Slovakia also used German 52s, and had a total of 29 on hire in January 1945.

At the end of the war there were approximately 600 class 52 locomotives in Czechoslovakia, most of them brought back in the retreats from the east. The majority of the class were later taken to the USSR leaving approximately 180 on the re-formed Czechoslovakian state railways (CSD) where they became class 555.0 as detailed in table 5a. Not shown in table 5a are the provisional numbers 555.0500–0566 allocated to certain locomotives in 1945 / 1946.

Nine class 555 locomotives were altered to 1524mm gauge and worked at Cierna nad Tisou on the border with the Soviet Union. The first two were regauged in 1947 and the seven others followed in 1948–1952. They were numbered 555.0601–555.0609 until converted back to standard-gauge between 1956 and 1965. (555.0604 was the exception to this rule, see table 5b for further details.) Another class 52 engine given a non-standard number was the condensing locomotive 52 1965 which was taken over by the Czechoslovakian army after the war and given the number 559.0701. It was transferred to the CSD in 1956, reconstructed as a conventional locomotive and given the number 555.0177.

In 1959 twenty class 555 locomotives were exported to Bulgaria and two more were sold to Jugoslavia in 1960 for industrial use [1]. However in 1962 and 1963 the CSD imported 100 additional TEs from the Soviet Union and these were given the numbers 555.0201–555.0300.

A major modernisation program for the CSD's 555s took place 1963–1966 when approximately two hundred locomotives were converted to oil-firing. The first engines were converted in Czechoslovakian works (Bratislava East, Most and Nymburk) but many others were processed in Kovel and Ivano Frankovsk in the USSR. Locomotives converted to oil-firing were renumbered from 555.xxx to 555.3xxx with the last 3 digits of their running numbers remaining unchanged.

The record of the 555.3s was marred by boiler explosions of three locomotives in the years 1964, 1965 and 1966 (555.3277 / 3280 / 3231 respectively). Nevertheless the class continued to work until the early 1970s, with the largest numbers of locos working from Bratislava East depot and Zdice.

Widespread withdrawals of the 555.0s and 555.3s started in the late 1960s and the last example of each class was retired in 1973. About fifty of these locos finished their working lives as heating units and were renumbered accordingly, so for example 555 3226 became K264 in 1966 and 555 010 became K558 in 1973. Four locomotives are known to have been saved for preservation, 555.3008 and 555.3254 in Slovakia, 555.3221 in the Czech Republic and 52 7620 (555.0153) in Austria.

The last major type of steam locomotive built in Czechoslovakia was the class 556 2-10-0, of which 510 were built between 1952 and 1958. These thoroughly modern locos had a larger and higher-pressure boiler than the German 52s, but certain other features were clearly inherited from the class 50 and 52 designs, including the wheelbase spacings, driving wheel diameter and most obviously the Witte smoke deflectors. The 556s worked until the end of CSD steam traction in the early 1980s.

[1] Another source suggests that only 18 CSD 555.0s were taken over by the BDZ. The locos dispatched to Jugoslavia were 555.0103 (which became 33-501) and 555.0153 (which became 33-502).

Table 5a: CSD class 555 and their original numbers.

	1	2	3	4	5	6	7	8	9	10
555.001-010	52 6625	6635	6645	6650	6655?	6661	6663	7447	6670	6671
011-020	52 6672	6674	7437	7440	7450	7452	7453	7454	7456	7457
021-030	52 7458	7459	7460	7461?	7465	7466	7467	7469	7470	7471
031-040	52 7472	7474	7476	7477?	7478	7483	7484	7487	7488	7489
041-050	52 7490	7493	7494	7498	7501	7502?	7506	7507	7515	7516
051-060	52 7518	7519	7520	7521	7522	7527	6653	7491	778	1359
061-070	52 1548?	-	2495	6673?	2597	-	-	526?	-	-
071-080	52 7146	-	-	7760	-	6629?	4842	5522	-	2744
081-090	52 7297	2781	6363	6366	833	1596	1726	1736	6370	-
091-0100	52 3120	3424	-	5430	3524	6074	-	2219	-	5327
555.0101-110	52 7187	-	2464	2820	3822	4372	5281	2525	2531	5143
111-120	52 5194	253	-	-	150	3920	3394	5634	4829	147?
121-130	52 4841	6364	-	6361	7752	6367	1675	-	7773	1681
131-140	-	743	780?	783	1571?	2739	3268	5045?	5631?	6377?
141-150	52 6383?	6823	7021?	7141	7144?	7148?	7150	7152?	7153?	7183
151-160	52 7315?	7388	7620	7622	7764	-	555	-	3709	308
161-170	52 3451	341	331	316	2771	277	339	-	3306	-
171-180	-	-	-	-	-	-	1965	555	6661	1725
181-185	52 1726	-	-	3530	1549					
555.0201-210	52 5719	5637	826	3670	827	3418	828	6086	625	6818
211-220	52 8024	650	1568	5181	6927	3900	527	5145	6915	6241
221-230	52 7047	1434	7362	7371	2317	7066	5344	4885	3696	6909
231-240	52 6814	8010	2839	7572	2857	2201	5105	3246	800	3803
241-250	52 7198	7499	2504	5788	7142	1428	4948	1123	6824	824
251-260	52 073	1310	7751	7565	4913	1240	3502	6122	7625	5026
261-270	52 3960	3142	6833	2204	7407	5837	1246	2171	1467	2715
271-280	52 3544	1264	5038	2471	6324	5464	832	1336	7650	5216
281-290	52 1607	3400	1715	7414	5968	4525	7563	7621	5230	6957
291-300	52 5981	6648	2882	7619	5286	255	034	5543	5966	1875

Notes:

1) Unknown previous numbers are represented by a dash.

2) Between 1963 and 1966 approximately two hundred 555s were converted to oil firing and renumbered by adding 3000 to the locomotives' numbers, so for example 555.001 became 555.3001.

3) The number 555.0184 was never carried. Loco 52 3530 was renumbered as 555.0603 whilst working on the broad-gauge at Cierna nad Tisou (1948-1965), after which it was converted back to standard gauge, converted to oil firing and renumbered as 555.3184.

4) According to Soviet archives, twelve additional 52s were selected to go to Czechoslovakia, namely 2368, 2419, 3754, 3887, 4367, 5174, 5183, 5192, 5206, 6770, 7360, 7523. (Order number TsT-14 of 26.4.63). It is not clear what happened to these locos.

5) 555.0170–0176 were assembled in Czechoslovakian workshops from parts of unserviceable 52s.

6) Many thanks are offered to Reimar Holzinger who supplied most of the data for locos 555.001–0177 (with confirmation coming from other sources).

Czechoslovakia: CSD class 555 locomotive heads a freight train out of Prague on the Decin line, June 1969. Photo Brian Stephenson.

Table 5b: Broad-gauge CSD 555s

Original Number	555.06 Number	Year to Broad-gauge	Year from Broad-gauge	Later Number
52 1549	555.0601	1947	1964	555.0185
52 1725	555.0602	1947	1961	555.0180
52 3530	555.0603	1948	1965	555.3184
52 3780	555.0604	1948		(1950 → USSR TE-3780)
555.087?	555.0605	1949	1961	555.0181
52 6235	555.0606	1950	1964	555.0183
555.0111	555.0607	1950	1964	555.0182
555.0157	555.0608	1952	1957	555.0178
555.006	555.0609	1952	1956	555.0179

see: Jan Koutny, Sirokorozchodné Lokomotivy u CSD, Zeleznicar 23 / 1987

CSD 555.3240 (ex 52 3803) on a Prague–Decin line freight at Podbaba near Prague, June 1969. Photo Brian Stephenson.

Newly-overhauled CSD 555.030 at Kolin shed, 23rd May 1968. Note the additional side-window on the cab. Photo Brian Garvin.

PKP Ty42-87 was built by Chrzanów in 1945. Here she heads away from Wolsztyn with a heavy freight train on 26th February 1992. Photo I. C. Simpson.

PKP Ty2-125 on a local freight south of Wroclaw in May 1983. This has the raised running-plate characteristic of former condensing locomotives. Photo P.J.S.

Poland

Poland inherited about 1200 class 52 locomotives at the end of the war and for the next forty-five years these remained the most numerous class of steam locomotive on the PKP. The Cegielski and Chrzanów factories completed some additional locomotives 1945–1946 and these were numbered as Ty42-1 to Ty42-150 though some at first carried their intended Reichsbahn 52.xxxx identities. The former German 52s were renumbered in 1947 as Ty2-1 to Ty2-1169 with further locos subsequently added up to Ty2-1206.

The postwar PKP had a shortage of modern freight locomotives and eventually took delivery of over 1000 new 2-10-0s of various types. With a clear long-term future, the ex-German 52s were modified to replace the most troublesome wartime features. Improvements included fitting adjusting wedges for the axle-boxes (where not already fitted), adding pressure-relief valves for the cylinders, shielding the leading axle against dirt falling from the smokebox, and modifications to the ash-pans to avoid starting fires on the track. Other improvements included the addition of smoke-deflectors, strengthening the bottom parts of Wannen tenders which were prone to cracking, and providing a lever-operated system for opening and closing water inlet covers without the need for the crew to climb onto the tender. The above changes are listed by Jan Piwowonski in his book "Porowozy Kolei Polskich": naturally some of these changes were also made in other countries where 52s were used, particularly the fitting of adjusting wedges and pressure-relief valves.

In the years 1962 - 1963 the PKP acquired two hundred additional Ty2s from the USSR, becoming Ty2-1207–Ty2-1406, and twenty more went to work on

PKP Ty2-1392 (ex 52 5044) with a Pt47 tender at Nysa depot in 1983. Photo P.J.S.

PKP Ty2-390 at Gorzuchów Klodzki on an afternoon freight from Klodzko to Walbrzych, 26th February 1985. Photo Keith Wilde.

Polish industrial lines retaining their German / Soviet numbers. The highest numbered PKP Ty2 is 1407 whose origins are not clear. It was first recorded in Bialystok in 1964, and is thought to be former 52 163.

In the mid 1960s about seventy Ty2s were converted to oil firing but little is known about these locomotives. One of them (Ty2-128) exploded at Bialystok on the 18th March 1967.

Other features were tried out experimentally on Ty2s, for example one was tested with a triple chimney in the 1960s and Ty2-1285 together with Ty2-953 received "gas-producer" fireboxes in the mid-1980s.

The Ty2s were used on almost all parts of the PKP network on both freight and passenger services. As in other countries their 80kmh top speed and light axle-loading aided their survival on secondary lines even after the takeover by diesel and electric traction on main-line duties. The most well-known depots still regularly using Ty2s in 1990 were Wolsztyn, Rzepin, Zbaszynek, Koscierzyna, Korsze and Elk. By June 1992 there were still Ty2s scattered around Poland derelict, in reserve or on heating duties, but Wolsztyn was now the last Polish depot working Ty2s and Ty42s in everyday service.

Locomotive identities

Locomotive historians have had considerable difficulty trying to identify the former German numbers of Polish Ty2 locomotives, particularly as no official renumbering list can be found in Poland. Details have been collected from Ty2 works plates and from the log-books of locos and their boilers, and the results clearly show a renumbering plan in which many locos were grouped by their year of manufacture (table 6). Within the groups Ty2-14–88, Ty2-89–838 and Ty2-839–1169 the majority of locos are numbered strictly in the order of the old 52 numbers. However, for reasons that are not at all clear, about thirty-five percent of locomotives between Ty2-14

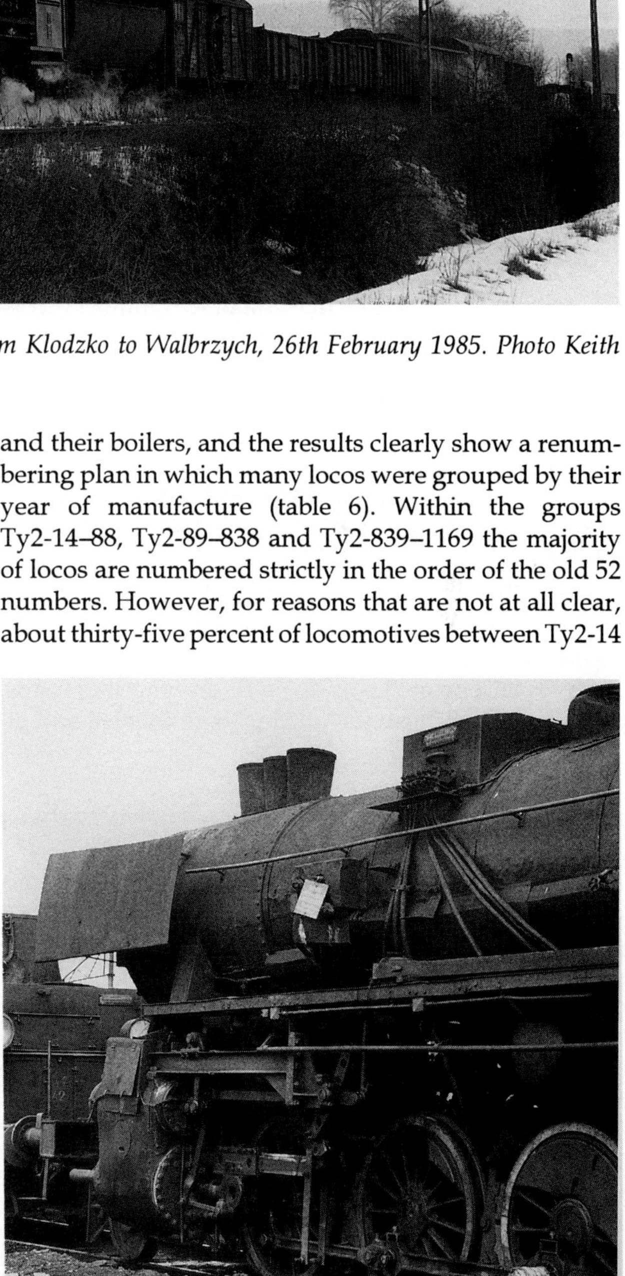

PKP Ty2 experimentally fitted with a triple chimney. Photo Brian Garvin.

PKP Ty2-406 on a twice-weekly local freight train on the Slawa branch, south of Wolsztyn, 25th March 1993. Photo I. C. Simpson.

and Ty2-1169 do not appear to conform to any renumbering plan, as shown by a survey of Ty2s still surviving 1983–1993. Further details are given in Appendix B.

The history of the 1945–1946 Chrzanów and Cegielski locomotives also presents a challenging task. It is commonly believed that the PKP had only 144 class Ty42 locomotives although the highest number was Ty42-150. It is not clear what happened to the "missing" locomotives, possibly they were taken out of Poland when almost new. One probable example is 52.1505 (the second loco with that number on the DR) whose official documentation describes the loco as Cegielski 887 and gives its boiler number as DWM 2814/1945.

It is possible to deduce the works numbers of Cegielski's Ty42s from the locomotives' works plates and from factory numbers found in loco and boiler log books. (See Appendix A). Some questions still remain unanswered, such as the fate of Cegielski's 894 which might be either Ty42-43 or Soviet TE-4578. Both locomotives were examined by enthusiasts (at Kluczbork in 1991 and at Rava Russkaya in 1992) but neither carried a chassis works plate.

The works-numbers of Chrzanów's Ty42s remain unclear because no works plates have been found on any of the locomotives, and there is also scant information in the loco log-books. The uncertainty about how many Chrzanów kriegsloks were completed 1945–1946 is one reason why it is not possible to give a definite total for the number of "class 52" locomotives finally constructed.

Table 6. Outline of Ty2 Renumbering

Polish Numbers	*Built*	*Description*
Ty2-1–Ty2-13	1943-44	ex-DRB Condensing locos *)
Ty2-14–Ty2-88	1942 *)	ex-DRB
Ty2-89–Ty2-838	1943 *)	ex-DRB
Ty2-839–Ty2-1169	1944 *)	ex-DRB
Ty2-1170–Ty2-1206		Various origins
Ty2-1207–Ty2-1406	1943-45	ex USSR 1962-63
Ty2-1407	1943	Origin unclear

* with some exceptions

PKP Ty42-87 (Chrzanów 1945) awaits departure from Wolsztyn with a freight for Zbaszyn on 26th February 1992. Photo I. C. Simpson.

CFR 150.1008 at Subcetate, 11th May 1971. This was one of 20 bar-framed locos supplied to Rumania by the Wiener Lokomotivfabrik in 1943. Like many other Rumanian locos it has been converted to dual coal-oil firing, and has an oil reservoir behind the coal bunker on the tender. Photo John Dawes.

Rumania

On the Rumanian state railways (CFR) it was the Prussian type G10 which became firmly established as the most numerous class of steam locomotive. Considerable numbers were bought new from Germany starting 1921 / 1922 onward, and these were later joined by second-hand DR locos and more than 500 examples built until the early 1940s by the Rumanian manufacturers Resita and Malaxa. During the years 1940 - 1944 Rumania hired further locomotives of this type from the Reichsbahn, followed by DRB classes 50 and 52.

During 1943–1944 the CFR took delivery of a hundred class 150.1 locomotives (German class 52) from the manufacturers shown below. The locos had bar frames and many had type K2'2'T26 tenders (similar to those used on the class 50), though the later examples had the austerity Wannen tenders.

150.1001-1020	WLF	16144-16163
150.1021-1030	Schk	12391-12400
150.1031-1040	MBA	13884-13893
150.1041-1043	MBA	13908-13910
150.1044-1060	Škoda	1444-1460
150.1061-1100	Hen	28070-28109

In the summer of 1940 Rumania lost about forty per cent of her territory to neighbouring countries. The following year she joined Germany in the war against the Soviet Union, regaining the territory of Bessarabia and taking part in the fighting as far east as Stalingrad. The terms of Rumania's armistice with the USSR (signed in September 1944) specified extensive reparation payments including large quantities of railway material, for example virtually the entire CFR class 140.4, but curiously the almost new class 150.1 locomotives were not included [1].

Some time later (dates uncertain) a small number of former German locomotives were taken into CFR ownership, including 150.1101–150.1123 which were examples of DR classes 50 and 52. Chassis works numbers show the identities of several locos, such as 150.1103 (52 4511) and 150.1123 (52 7468). The former loco had been taken by the Soviets from Hungary shortly after the war, and the latter loco had been taken by the Soviets from Austria in 1948. It is known that 150.1114 and 150.1115 were ex DR class 50 locomotives. See Appendix F.

In the early post-war years Rumania commenced building the CFR class 150.0 from the designs of the German class 50. These were built until 1960 with the highest CFR number being 150.282. Ten locos of this type were exported to China where they became class DK5 (see Far East).

The 150.1s and later the 150.0s were the most modern steam locomotives used until the advent of diesels. They worked both passenger and freight services, frequently working in multiple, and indeed freight trains with as many as four 2-10-0s could be seen in the Turnu Severin area.

Both types of 150s were a familiar sight working trains until the early 1970s and a few were still in use in 1985. Lok-Report 7/86 describes the use of 150.1s for shunting and occasional local freight trains in the area around Cîmpulung Moldovenesc (northern Rumania). Many steam locomotives remained stored or dumped at depots in the early 1990s, including Prussian designs P8 and G10 and the CFR classes 150.0 and 150.1. At that time a few G10s were still in use for shunting duties, so it was this eighty year old design which outlasted all other types in regular service on the CFR.

[1] Rolling-stock taken by the Soviets from Rumania 1944–1945 included locomotives formerly on hire from Germany. According to a report from the Allied Control Commission for Rumania, fifty-seven of these locomotives were taken away by the Soviets by June 1945, with the following intended destinations:

15 to Jugoslavia,
25 to Bulgaria,
10 for the USSR (Bessarabia / Bukovina)
7 for the USSR.

The same report suggests the number of hired German locomotives in Rumania at the end of the war was 137. The types of the locomotives are not specified.

Jugoslavia

In 1944 the railways of the wartime state of Serbia took delivery of fifteen new "52s" and numbered them 33-001 to 33-015. Similarly the wartime state of Croatia had 24 locos of this type numbered 30-001 to 30-024. After the liberation of Jugoslavia these locos were allocated the numbers 33-001 to 33-039 by the JDZ (Jugoslav State Railways). With the exception of 33-040, of which nothing is known, numbers up to 33-177 were allocated to class 52 locomotives left behind by German forces withdrawing from the Balkans.

War-ravaged Jugoslavia had a severe shortage of steam locomotives in 1945. In response the United Nations (UNRRA) delivered more than 50 American-built S160 locos and 65 Liberation 2-8-0s. In 1947 and 1948 fifty more 52s were transferred from Soviet ownership, more than half being "Trophy" locomotives taken from Austria. These were later supplemented by thirty-five 52s which arrived from West Germany in 1952. The last main batches of 52s to come to Jugoslavia were 76 from the USSR which arrived in 1964 and 1966 as shown in table 7 below. Finally it should be mentioned that five industrial 33s worked at Kreka coal mine, namely 33-501 and 33-502 which came from Czechoslovakia in 1960, plus 33-503 to 33-505 which came from the Soviet Union in 1964.

Jugoslavian Railways Kriegslok at Belgrade on an express passenger working in 1966. The loco had recently been acquired from the Soviet Union and is still carrying its Soviet number 52 320. Later it became JZ 33-330. Photo Dr P. Ransome-Wallis, National Railway Museum, York.

Jugoslavian Railways 33-054 (ex 52 702) at Skopje, 27th August 1961. Photo D. Trevor Rowe.

The 33s were the most numerous type of steam locomotive on the JZ (as the Jugoslavian railways were called from about 1953 onward). They could be seen on most parts of the railway system and served on a wide range of duties, from heavy mainline freight traffic to passenger duties on secondary lines. They were one of the JZ's last steam types and were still in use in 1985, for instance in the Lapovo, Novi Sad and Ruma areas.

In Bosnia some use was made of steam locomotives during the 1992-1995 war, including 33-236, which was observed on a passenger train at Tuzla in March 1993. A report in the Times (4th May 1996) refers to eight class 33 locomotives at Kreka coal mine near Tuzla. These were brought out of storage to replace diesels on coal trains in 1993, and some of them continued in service after the end of fighting in 1995.

Locomotives in the Tuzla area in 1996: 33-064, 33-216, 33-236, 33-248, 33-501, 33-503, 33-504, 33-505.

Table 7 - Summary of origins of class 33 locomotives

JZ 33-001 - 33-015	ex SDZ 33-001 - 33-015
JZ 33-016 - 33-039	ex HDZ 30-001 - 30-024
JZ 33-040	?
JZ 33-041 - 33-177	ex DRB class 52, circa 1945
JZ 33-178 - 33-179	ex DRB class 50
JZ 33-180 - 33-217	ex DRB class 52, from Soviet forces 1947
JZ 33-218 - 33-229	ex DRB class 52, from Soviet forces 1948
JZ 33-230	ex DRB class 50, from Soviet forces 1948
JZ 33-231 - 33-265	ex DB class 52, 1952
JZ 33-266 - 33-320	ex USSR, 1964
JZ 33-321 - 33-341	ex USSR, 1966
33-501 - 33-502	ex CSD, 1960 ex 52 2464, 52 7620
33-503 - 33-505	ex USSR, 1964 (TE 1345, 793, 3654)

BDZ 15-247 near Kaspican, 19th May 1965. Photo D. Trevor Rowe.

Bulgaria

Bulgaria was one of the Axis powers during the second world war, until shortly before the occupation of the country by the Red Army in September 1944. At that time Bulgaria had many Reichsbahn locomotives on loan including 125 of the class 52. After continuing to hire Reichsbahn locos from the Soviets, eighty-five 52s were purchased circa 1946 by the Bulgarian State Railways (BDZ) and renumbered as their class 15 [1].

In the 1950s and 1960s further additions to the class 15 fleet came from East Germany (30 locomotives), Czechoslovakia (20) and the USSR (140) as shown

BDZ 15.82 and 15.67 on a passenger working near Sofia, May 1965. Photo D. Trevor Rowe.

TCDD 56520 (ex DRB 52 4859) and 56502 at Inay in 1984, with a mixed train from Alasehir to Usak. The leading locomotive retains a chimney cowl and insulated tender, two of the "extra frost-protection" features that were applied to some of the early class 52 locomotives. Photo P. J. S.

below. The final class total was 275 locomotives, the most numerous steam type in Bulgaria.

15.01–15.85	ex-DRB, in Bulgaria at the end of the second world war
15.86–15.105	ex-DR 1956
15.106–15.125	ex-CSD 1959
15.126–15.135	ex-DR 1959 (initially on hire; purchased 1961)
15.136–15.275	ex-USSR 1961-1964.

The BDZ 15s were widely used on most types of traffic. Not surprisingly in this mountainous country, double-heading of 15s on freight trains was commonplace, and pairs of 15s were sometimes also seen on passenger workings. Like certain other BDZ classes, the majority of 15s were equipped with dual coal / oil firing. When locomotives were working under very easy conditions firing was by coal only, but fuel-oil could be injected into the firebox when a higher steaming rate was required.

The 15s survived until the last days of regular steam working on the BDZ in 1980. Thereafter some were used as heating locomotives, and many of the others were stored until about 1993 at Asenovo, Bodrovo, Kaloyanovetz and various other "reserves". Visits to the reserves by several enterprising enthusiasts revealed the former German 52.xxxx numbers of some of these locomotives.

[1] Of the other hired 52s, one (52 4045) went to the USSR via Hungary, and 39 are believed to have been transferred to Jugoslavia, including 28 in the Skopje area. See "Die Lokomotiven der Bulgarischen Staatsbahnen" by Dimiter Dejanow, published by Slezak in 1990.

Turkey

A total of fifty-three class 52 locomotives were delivered to neutral Turkey 1943–1944. Forty-three of these were Reichsbahn locos and were initially on loan. They were later handed over to the TCDD (Turkish state railways) becoming 56511–56553, whereas numbers 56501–56510 were allocated to ten 52s delivered to the TCDD directly from the manufacturer. Most of the 565xx worked on the Asian side of the Bosporus, though a small number of the class worked in European Turkey and operated trains to the Bulgarian and Greek frontiers, including the Istanbul Express. Together with other steam types they were still working from Halkali and Çerkezköy in 1972. Not long afterwards these 565xx were replaced by diesels and transferred to Anatolia.

The hills and mountains of Asian Turkey ensured plenty of hard work for the 565xx. Double heading was commonplace, for example on the route Izmir–Usak–Afyon where locos with heavier axle loadings were not permitted. Three and sometimes four 565xx could also be seen struggling with heavy freights on the line from Balikesir to Bandirma. On this route work was shared with the class 462xx built 1942 in the USA for the British War Department, so here it was possible to see both German and American-built "Kriegsloks" working together on the same train.

Depots with class 565xx locomotives in 1975 included Afyon, Balikesir, Bandirma, Eregli, Erzurum, Izmir, Kars, Konya, Tavsanli and Ulukisla. The locos at Erzurum were used mainly for suburban passenger duties, and further east the 565xx were used on trains from from Kars to the Soviet border. By 1985 the Balikesir 565xx had largely been replaced by diesels, and in turn the 565xx were replacing the old Prussian G8 and G10 types on the Burdur, Isparta and Egridir branches. Regular line working of the 565xx and other TCDD steam locomotives ceased in 1990. About ten 565xx were retained for operating special trains, and three are known to have been preserved as static exhibits, namely 56504 at Ankara, 56523 at Camlik and 56547 at Usak.

TCDD class 565xx locomotives

TCDD	Builder	Works number	Year	DRB
56501-56510	Henschel	27735-27744	1943	
56511-56515	Borsig	15461-15465	1943	52 364–52 368
56516-56517	MBA	13918-13919	1943	52 4855–52 4856
56518-56525	MBA	13921-13928	1943	52 4857–52 4864
56526-56527	Schwartzkopff	12503-12504	1943	52 6062–52 6063
56528-56535	Schwartzkopff	12507-12514	1943	52 6066–52 6073
56536-56543	WLF	16738-16745	1943	52 7285–52 7292
56544-56553	WLF	16878-16887	1943-44	52 7425–52 7434

TCDD 56522 near Yeniköy on the 7:30 freight from Balikesir to Bandirma, 4th October 1983. 56553 and 56165 are assisting at the rear of the train. Photo E. Talbot.

TCDD 56522 and 56529 leaving Usak on the 13:00 mixed train to Afyon, 24th December 1985. Photo E. Talbot.

TCDD 56535 (ex 52 6073) leaving Yildirimkemal on the 7:10 from Afyon to Usak, on 25th October 1984. Photo E. Talbot.

TCDD 56522 on an international train at Pithion in Greece, 21st May 1965. Photo D. Trevor Rowe.

USSR

Class 52 locomotives first came into Soviet hands in 1943, and official records show at least ten on the Stalin railway by 1st January 1944. Captured German locomotives were used by military railway staff on the 1435mm gauge railways in the area of the battle front, and a number (including 52s) later returned to Europe with the oncoming armies. One of many locos remaining in the Soviet Union was 52 215, which in 1944 was taken to the locomotive testing establishment at Shcherbinka near Moscow.

After the war there was a great "migration" of class 52 locomotives to the USSR as reparations. These so-called Trofeinyi ("Trophy") locomotives came from Germany 1945-1947 and also from Austria 1945-1953, Czechoslovakia, Hungary and Rumania. Ultimately there were more than 2130 of this type in the post-war Soviet Union, clearly exceeding the number in any other country.

The total of Soviet 52s cited above includes not only the locomotives of Soviet Railways but also those in industrial use and others which were badly damaged and written-off 1945-1951. The total does not include captured locos which in 1944 and early 1945 followed the Soviet forces back to Poland and Germany and did not return to the USSR. [1]

The process of converting 52s to 1524mm gauge involved, in many cases, the use of mobile workforces who made temporary adjustments to the wheel rims. The regauging work was completed at the workshops at Unecha, Kaliningrad (Königsberg), Vilnius, Liepaja, Daugavpils, Tallinn, Stanislav (Ivano Frankovsk) and Lvov. Similar work was also undertaken at well equipped depots such as Shchors, Kishinev, Chernyakhovsk (Insterburg) and others. Fully regauged 52s had their wheels moved outward on their axles, thus leaving a distinct "dimple" at the centre of each wheel which makes the 1524mm locos easy to identify.

In the late 1940s the conversion of 52s to broad-gauge was also carried out in Czechoslovakia and Rumania. Locomotives which were given major repairs 1950-1951 at Bucharest Grivita acquired weighty cast-iron number plates, as shown in the photograph of 52 3505. [2]

In 1945 the warn-torn railways in the western USSR were in need of extensive reconstruction. The "locomotive cemeteries" contained large numbers of 52s and other German types, and many of these were later repaired in the works at Stanislav (Ivano Frankovsk). Among the other works doing overhauls was Daugavpils which in November 1948 repaired and regauged 52 5694 now preserved at St Petersburg. It is probably at Daugavpils that this loco received its present chassis from 52 422 which was previously recorded in 1946 in the Novocherkassk area.

Over the years the less satisfactory wartime features of the 52s were replaced. For example, the the poor freewheeling performance of the 52s was improved by fitting Trofimov piston valves.

Standard features of the Soviet 52s included the rebuilt front-end with smaller smoke-box door, smoke-box-drainage pipe, and the new off-take valve behind the chimney. The list of non-standard additions is almost endless and includes the popular clerestory cab-roof, a double size sandbox and lengthened chimney.

One interesting experiment tried out on a 52 to raise its pulling power was the provision of variable axle-loading. In this system the weight on the leading axle could be relieved under pneumatic control, and the axle-loading on the driving wheels was correspondingly increased.

In accordance with order B-60570 of 23.8.1952 the 52s were officially classified TE, where T (Trofeinyi) meant "captured" and the designation E showed the same power category as the Soviet type "E". Other ex-German locos were reclassified on the same principle, so the 03 Pacifics became the TS, the 38s became the TA, the class 42 became the TL and the class 50 became the TYe. [3] In

Soviet 52 860 at Riga Zasulauks in the late 1940s. Collection of the Latvian Railway Museum.

Standard-gauge 52s stored at Vysokoye Litovsk (Byelorussia) 26th April 1991. Photo A. Gogolin.

almost all cases the German numbers were left unchanged.

The TEs worked in the western parts of the Soviet Railways (SZD) network and were allocated to over 120 depots (see table 8a). They proved to be a useful and generally popular type commonly pulling freight trains of 1500 - 2000 tonnes and more, and they also saw use on passenger services. The livery of freight locomotives was usually black with white lining and white lettering with red wheels and red chassis, though from 1956 the colour of bar-frames had to be black according to the new overhaul regulations. TE locomotives used for passenger services were mostly painted black or green but some were pale blue. A few TEs carried the names of famous Soviet people, and these were particularly colourful and brightly decorated.

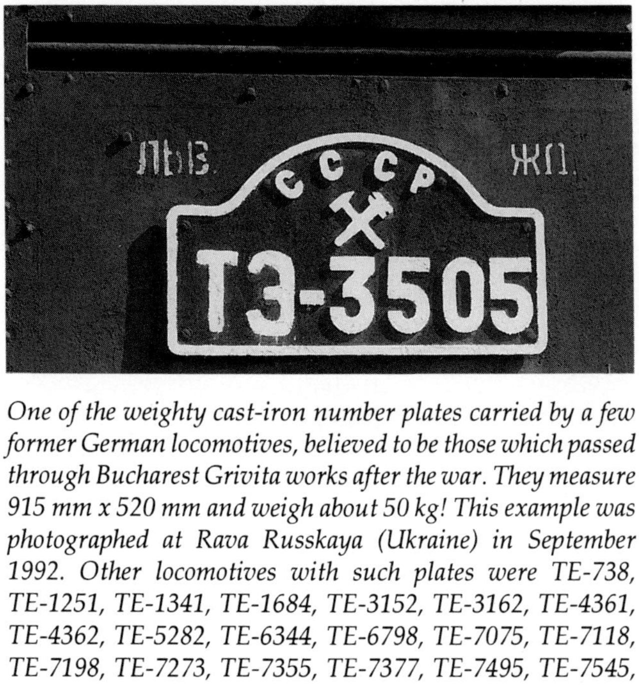

One of the weighty cast-iron number plates carried by a few former German locomotives, believed to be those which passed through Bucharest Grivita works after the war. They measure 915 mm x 520 mm and weigh about 50 kg! This example was photographed at Rava Russkaya (Ukraine) in September 1992. Other locomotives with such plates were TE-738, TE-1251, TE-1341, TE-1684, TE-3152, TE-3162, TE-4361, TE-4362, TE-5282, TE-6344, TE-6798, TE-7075, TE-7118, TE-7198, TE-7273, TE-7355, TE-7377, TE-7495, TE-7545, TE-7571, TE-7648, TE-7783. Photo P.J.S.

Now a few words about the 52s built with condensing equipment. This type first came into Soviet hands in the spring of 1944, and there were eventually at least 21 in Soviet ownership. Three of these "camels" (52 1863, 1872 and 1902) quietly lived out their days in Odessa, where they were used as heating units at Sortirovochnaya from 1945 until 1950. On 18th September 1950 they were transferred to a different railway department and there we lose trace of them. The most widely travelled of the condensing locos seems to have been 52 1933 first recorded in the Vinnitsa area on 1st April 1944. This was allocated to the Omsk Railway between 1948 and 1952 before returning to the Odessa–Kishinev Railway and final withdrawal in 1956. Some of the locomotives had their condensing equipment removed and gave many more years service, including TE 1856, 1875, 1888, 1936 and 1937. TE 1937 worked for a long time shunting at Kotovsk (Odessa Railway) and finished its working days as yard pilot at Gulbene (Latvia) before being

Table 8a: Depots of the class 52 in the former Soviet Union.

Although more than 120 depots are known, the list is incomplete.

Russia: Moscow Railway
Fayansovaya
Lgov
Roslavl
Smolensk
Unecha

Russia: October Railway
Bologoye
Dno
Gatchina
Khovrino
Khvoynaya
Leningrad-Baltisk
Leningrad-Finland
Leningrad-Vitebsk
Leningrad-Warsaw
Luga
Medvedevo
Novgorod
Novosokolniki
Ostashkov
Pskov
Rzhev
Sonkovo
Torzhok
Velikiye Luki

Russia: Pribaltic Railway
Chernyakhovsk
Kaliningrad
Sovetsk

Byelorussia
Baranovichi
Bobruysk
Brest Central
Brest East (Vostok)
Gomel
Grodno
Kalinkovichi
Krichev
Krulevshchizna
Lida
Luninets
Mogilev
Molodechno
Osipovichi
Polotsk
Vitebsk
Volkovysk
Zhlobin

Ukraine: Southwest Railway
Chernigov
Fastov
Grechany
Kazatin-1
Kazatin-2
Kiev Moscovski
Korosten
Ovruch
Shchors
Shepetovka
Zhitomir
Zhmerinka

Ukraine: Lvov Railway
Adynkata
Chernovtsy
Chop
Khodorov
Khyrov
Kivertsy
Kolomyya
Korolevo
Kovel
Lvov East (Vostok)
Lvov West (Zapad.)
Mostiska
Mukachevo
Rava-Russkaya
Sambor
Sarny
Stoyanov
Stryy
Ternopol
Vladimir-Volynski
Zdolbunov

Ukraine: Odessa Railway
Artsyz
Dolinskaya
Golta
Kherson
Khristinovka
Kirovograd
Kotovsk
Nikolayev
Odessa Freight Depot (Tovarnaya)
Odessa Marshalling Yards (Sortirovochnaya)
Pomoshnaya
Shevchenko
Tsvetkovo
Znamenka

Ukraine: Southern Railway
Grebenka
Poltava
Priluki

Moldavia
Bendery
Bessarabskaya
Beltsy
Kishinev
Oknitsa
Reni
Sholdaneshty
Ungeny

Estonia
Tallinn
Tapa
Tartu
Valga

Latvia
Daugavpils
Gulbene
Jelgava
Liepaja
Rezekne
Riga Ciekurkalns
Riga Skirotava
Ventspils

Lithuania
Kaunas
Klaipeda
Radviliskis
Siauliai
Svencioneliai
Vilnius

TE-5314 in steam at the standard-gauge reserve at Vysokoye Litovsk, 26th April 1991. Photo A. Gogolin.

Stored standard-gauge locos at Vysokoye Litovsk, 26th April 1991. Photo A. Gogolin.

TE-6432 undergoing maintenance at Brest East, together with a Soviet type ER locomotive, 10th July 1989. Photo A. Gogolin.

Soviet 52 5330 at Liepaja (Latvia) in 1958, not yet with the TE designation on the cabside. Collection of Toms Altbergs.

withdrawn in 1987. TE 1936 has been stored for many years on the Byelorussian Railway and was still in existence in 1994.

It might seem surprising that the condensing 52s did not see wider use on Soviet railways, but it should be pointed out that by the end of the 1940s less than half of the Soviet-built SO-19 condensing locomotives were still active, and these were only working on secondary duties.

Although almost all "Kriegsloks" worked in the USSR under their original German numbers, there were more than forty exceptions to this rule. For example, TE 4566, 4567 and 4578 which were built in Poland in 1945. A few locos carried the wrong number for reasons which have been forgotten, such as 52 2821 which was numbered TE-2021. In addition, records are clear that there were two locos with the number TE-1447, two with the number TE-2534, and two with the number TE-6944. However, the best known locos with non-German numbers were TE-8001-8034/8036 and TE-9101-9103.

The 8001 series originated from the locomotive works at Stanislav (Ivano Frankovsk), where a large number of 52s had been brought for repair and rebuilding. The availablity of spare parts as well as totally wrecked 52s prompted the decision to produce a single locomotive out of two or three wrecks. The first such rebuild was completed in 1949 and was assigned the ministry inventory number 8001. It had a bar-framed chassis but subsequent locos were constructed on both bar and plate frames without any apparent system. Some of them had a distinctive rectangular casing for the steam dome.

The prototype TE-8001 deserves special mention. In September 1952 it was transferred to the Gaino-Kaiskaya railway which is the main timber-carrying railway in the Kirov region. This complex system starts at Verkhnekamskaya and extends northwards to Syktyvkar. For a long time exiles live in this area, and nowadays convicts work on timber felling. The place is known as "Vyatlag", short for the Vyatka camps. In company with twenty other TE class locomotives, TE-8001 hauled trains of 2000 tonnes and more through the taiga. The so-called "moustaches", the temporary branch lines for carrying timber out of the forests, functioned only for a year or so until the supplies were exhausted. TE-8001 worked from Lesnaya depot and was written-off in the first half of the 1970s.

Soviet TE-3901 with double-length chimney at Jelgava depot (Latvia) circa 1955. Collection of the Latvian Railway Museum.

TE-107 near Tapa in Estonia in the late 1950s. Photographer unknown.

The second locomotive, TE-8002, had plate frames. It was allocated to Sarny depot and then to Odessa Tovarnaya where it was used on trains to Razdelnaya. Gradients were such that the normal load to Razdelnaya was 1700 tonnes, but on return journeys drivers sometimes took as much as 4000 tonnes! The last locomotive in the series was TE-8036 which had bar frames and worked from Lvov East depot where it was used for heavy shunting.

The enigmatic 9101 series comprised just three locomotives. No definite information is known about their origins, apart from the earliest records of their use. TE-9101 was first recorded in Latvia in February 1950, TE-9102 in Latvia in January 1950 and TE-9103 on the Lvov Railway in October 1949. Further details of the TE8001-series and the TE9101-series are shown in Appendix D.

TE locomotives were widely used for a period of ten years or so. In the national development plan 1956–1960 diesel and electric traction spread through the busiest parts of the SZD, and cascaded steam locomotives such as the class L replaced many of the less powerful TE class. Many displaced TEs were overhauled and put into storage at "cold depots" such as Zhlobin North (about 40 TEs in 1963) and Luninets (approximately 75 TEs in 1964). During the period 1962–1966 about seven hun-

TE-1512 leaves Tallinn with a heavy freight train some time in the mid 1950s. Photographer unknown.

TE-6962 at Valga in Estonia. Date and photographer unknown.

dred TEs returned to the railways of eastern Europe, namely:

Germany	1962	60 locos
Czechoslovakia	1962-1963	100 locos
Hungary	1963	100 locos
Poland	1962-1963	220 locos
Bulgaria	1961-1964	140 locos
Jugoslavia	1964, 1966	79 locos

Whilst many TEs were being replaced in the late 1950s others were starting a new era as oil-burners. Among the first of these was TE-7162 (converted in 1956) and another was TE-7354 which was converted to shale-oil firing at Tapa depot in 1957. Oil firing for TEs became widespread on the SZD in the early 1960s. A number of Czechoslovakian 52s (CSD class 555) were similarly converted at Kovel and Ivano Frankovsk on the Lvov Railway.

One of the least-known parts of the TE story concerns the locomotives which worked in industry. Sixteen are recorded as being transferred to industrial use in 1945, and these were followed in 1946 by other locomotives which originated in eastern Germany. Eventually about 60 worked in the non-ferrous metal industry and 100 in the coal industry. Significant numbers also worked on forestry railways (in the Kirov, Gorki and Archangel areas) and in various other types of industry. A list of some of the industrial depots is shown in Table 8b.

TEs continued in traffic on the SZD until the 1970s. In 1972 they could be seen, for example, at Kaunas on various local services, at Kaliningrad (local freight) and at Tapa and Narva in Estonia. Over six hundred were still in existence in 1984 when the SZD introduced a computer numbering scheme for its rolling stock. The TEs became the class 1042 and were renumbered as 1042.001 to 1042.623. [4]

At the start of the 1990s the well-known Soviet locomotive reserves remained a home for about 300 mostly European-gauge TEs. A few broad-gauge TEs could also be found at various depots, such as heating locomotive TE-4564 which was observed in steam at Chernyakhovsk in August 1993. It is still possible to see TEs in the Russian Federation as a number have been preserved for museums and railtours, such as TE-622 and 6744 now owned by Intertrack of Moscow.

[1] An example of Soviet 52s returning to Europe is provided by order 11000 Ts of the Peoples' Commissar (Railways), which dispatched the following fifteen locos from the Stalin and Lvov Railways to the front line in autumn 1944: 52 2166, 2267, 2268, 3391, 4809, 5307, 5672, 5791, 5794, 5935, 6048, 6244, 6247, 6993, 7396. Some of these locomotives are believed to have remained in Poland after the war, though not 52 5307 which had returned to the Soviet Union by October 1945. Locomotives that returned from the USSR to Germany included 52 701 and 52 7129.

[2] Workshops known to have been used for repairs and regauging work on reparation 52s include: Ustí nad Labem, Ceská Lípa and Ceská Trebová in Czechoslovakia, Bucharest Grivita and Iasi in Rumania.

TE-8033 at Radviliskis in Lithuania, 23rd September 1965. Note the rectangular casing for the steam dome and lengthened chimney. Photo Alexander Vassiliev.

TE-9102 at Lvov East depot in 1975. Photo A. Nikolski.

Oil-fired TE-1553 at Ostashkov on the October railway in September 1985. Photographer unknown.

TE-567 in industrial use on the Gaino-Kaiskaya railway, the main timber transporting line in the Kirov region. This photo dates from 13th August 1958 and shows the loco at Lesnaya after arrival with passenger train number 62 from the direction of Kirov. Coal and timber were used for firing, which explains the cage on the tender. Photo V. Belyaev.

[3] The Soviet classification TYe of the former German class 50 locomotives becomes **TE** in the Cyrillic alphabet.

[4] The ministerial order 22Ts of 15th May 1984 provided 8-digit computer numbers for all rolling-stock. The eighth digit is a "check" digit. More details about the new numbers are given in Appendix H. In general steam locomotives retained their original numbers on the front buffer-beam.

Table 8b: Class 52 Locomotives in Soviet Industry

Listed below are some of the locations where class 52 locomotives worked on Soviet industrial railways.

Bulanash (North-east of Ekaterinburg). Industrial railway for coal.
52 822, 1492, 4904
Kohtla-Järve (Estonia). Shale-oil industry.
52 102, 1115, 1558, 3207, 3390, 3408, 3522, 5466, 5530, 5641, 6291, 7354
Lesnaya (North-east of Kirov). Forestry railway.
52 561, 567, 749, 1150, 1349, 1688, 2547, 2684, 3414, 3831, 4766, 4914, 5010, 5054, 6030, 8001 (More than 30 locos).
Makeyevka (North-east of Donetsk), "Makeyevugol" industrial railway for coal.
52 5193, 7477 and 7 locos of type TL (DRB 42)
Puksoozero (Arkhangelsk district). Forestry railway.
52 706, 1650, 7196 (About 5 locos).
Revda (West of Ekaterinburg). Copper industry.
52 303, 536, 1577, 2611 and others.
Salavat (South of Ufa in the Bashkir republic)
2 locos until about 1954.
Sukhobezvodnoye (Gorki district). Forestry railway.
Approximately 20 locos until about 1956, eg 52 7125.
Uglegorsk (Donbass). Coal industry.
52 337, 5115, 5819
Ural Aluminium Works. (Kamensk Uralski).
52 303, 1309, 1577, 5531, 7044

Estonia, Latvia, Lithuania, Russia, Belarus, Ukraine, Moldova (Successor states of the USSR)

The three Baltic republics, Estonia, Latvia and Lithuania used a considerable number of TE locomotives until the 1960s and early 1970s when they were displaced predo-

TE-1383 at Sovetsk (formerly Tilsit) in September 1965. Alexander Vassiliev is at the window.

Oil-fired TE-781 at Liepaja (Latvia) in 1978. Photo A. Nikolski.

minantly by diesel traction. More than a hundred were retained in reserve or as heating locomotives, but few survived to 1991 when the Soviet Union officially recognised the Baltic States as independent. Preserved locomotives include TE-3368 in Estonia and TE-021 and TE-036 in Latvia.

The newly independent Latvian railways inherited eight Soviet TEs in various states of repair (as well as ninety L class 2-10-0s). A visitor to Riga in 1992 found TE-6747 and TE-7528 being cut-up for scrap at Skirotava depot, whilst at Tornakalns TE-6761 was in use as a heating locomotive in the company of former heating loco TE-023. At Radviliskis depot in Lithuania visitors in 1992 saw two TEs out of use, namely TE-120 and TE-4853, and the nearby scrapyard contained TE-7231 and the remains of TE-6955.

Over the border in the Kaliningrad district of Russia, visitors in 1992 and 1993 found no less than 116 TEs, most of them in the standard-gauge reserves in Bagrationovsk, Krasnovka, Mamonovo and Zheleznodorozhnyj. The only TEs seen in steam were European-gauge TE-777 (used on a special train for visitors from Germany) and TE-4564 used as a heating locomotive at Chernyakhovsk container depot. See appendix E for more details.

After declaring independence from the Soviet Union in August 1991, Belarus had at least 106 of the class TE mostly stored at Vysokoye Litovsk (53), Berestovitsa (15) and at Bruzgi (8) near Grodno. Small numbers of Russian-gauge locos were stored at Luninets (12) and Neman (5). The most notable of the survivors were TE-8021, TE-8026 and ex-condensing locomotive TE-1936. See Appendix E.

About 120 TEs remained in the Ukraine after independence was declared in 1991. The largest concentration of these was probably at Rava Russkaya where in 1992 there were 39 on the European gauge and one on the Russian-gauge. Other locations of stored European-gauge TEs were Grodovitsa, Khyrov and a location near

Table 8c: Locomotive histories (Examples from locomotive log books)

TE-023

Delivered:	4/11/42
Boiler:	Henschel & Son 1942
Boiler Nr:	26941
Boiler register number:	60388

27-04-47	Regauged to 1524 mm
21-05-48	Gulbene
	Rezekne
13-04-49	Gulbene
12-02-52	Daugavpils
15-07-59	Riga
05-02-63	Liepaja

In 1992 TE-023 was at Riga Tornakalns in extremely poor condition. It had apparently been used as a stationary boiler.

TE-622

13-05-48	Khristinovka
17-05-49	Kolomyya
12-04-50	Sambor
07-05-51	Velikiye Luki
21-03-52	Torzhok
-	Kaliningrad
12-07-60	Chernyakhovsk
-12-60	Kaliningrad

TE-622 was brought from Kaliningrad to Moscow in January 1993 to join Intertrack's collection of preserved locomotives at Shcherbinka.

Ukraine: TE-5653 hauls a demonstration freight train for a group of railway photographers, 28th April 1993. The location is a bridge over the River Irsha between Korosten and Zhitomir. Photo Jack Hodgkinson.

Ukraine: Standard-gauge locos at Rava Russkaya in 1992. Nearest the camera is TE-7496, and behind it are TE-7545 and TE-1322. Photo P.J.S.

Mostiska. It is possible that some locos also remained at Vyerbka near Kovel, Korolevo (near western end of the Rumanian border) and at Porubnoye near Chernovtsy. See Appendix E.

Little was known about the railways in the newly independent republic of Moldova (formerly Moldavia) until the visit of the first western railway enthusiast in the summer of 1994. TE locomotives were found only at Ungeny where a total of fourteen standard-gauge examples were present. Sixteen years earlier, Alexander Vassiliev had visited Ungeny and had seen a total of forty TEs all stored as reserve locomotives.

Locos at Ungeny in 1994: TE-435, 656, 1370, 1392, 2188, 2394, 2879, 3951, 6007, 6675, 7176, 7345, 7395, 7438.

The Far East

According to former Soviet archives, nine TEs from Moldavia were selected in 1984 for delivery to Vietnam. They were: TE-1263, 1390, 2363, 2727, 5096, 5470, 5490, 5560, 5978.

It is also known that three Polish Kriegsloks were donated to Vietnam some time in the mid 1980s, namely Ty2-464, Ty2-1214 and Ty2-1220. Visitors to Vietnam have reported that these locos were not put into service but were stored at the following locations:

Former Soviet TE-1263 stored near Hanoi in Vietnam in 1989. Ten years earlier it was at Ungeny depot in the Soviet republic of Moldavia. Photo: J. A. de Koningh.

– Dong Anh (near Hanoi): the three Polish locos were seen here in 1987 together with TE-5978 and three others not identified.

– Yen Vien (near Hanoi): TE-1263 was stored here in 1989.

– Near Kep: Five Kriegsloks including TE-1390 and TE-2727 were found here in 1993.

Press reports suggest that the last of the Vietnamese 'Kriegsloks' was scrapped in 1994.

Also in the Far East is former ÖBB 52.2436 which is preserved in Japan. China has preserved a closely related locomotive, DK5-250 at the railway museum at Shenyang. The ten Chinese locos DK5-241 – DK5-250 were built by Resita in Rumania substantially to the design of the Rumanian class 150.0, and were derived ultimately from the German class 50.

Looking ahead

The class 52 clearly has a secure future with Europe's preserved steam railways and other museum collections. At the time of writing (Summer 1996) over two hundred have been acquired for preservation, a total which probably exceeds any other type of steam locomotive. Hopefully more examples will be saved from the surviving locomotives still stored in eastern Europe and the former USSR. A list of "museum locos" can be found at the end of this book.

List of class 52 locomotives

The following list of locos 52 001–52 7792 is divided into batches, each identified by manufacturer and month/year of delivery. Against each locomotive can be seen the post-war users and the date of withdrawal from service where appropriate. To make this book as compact as possible, extensive use has been made of abbreviations, which are explained below.

The list of locomotives has been copied directly from this book's German edition, and therefore contains a few German words and phrases which are explained on page 128. Place names are as written in German, for example with W instead of V and with J instead of Y.

Soviet locos are shown with their last-known SZD Railways together with year of transfer to each Railway. Where the year is underlined, this indicates not a transfer, but an observation or official record that the loco was on the Railway during the shown year. For example,

52 1933 USSR (Sw 47, Om 48, OK 54) + 56

shows that 52 1933 was recorded on the Sverdlovsk Railway in 1947, was transferred to the Omsk Railway in 1948, joined the Odessa–Kishinev railway in 1954, and was finally withdrawn in 1956.

It will be seen that most of the locomotives taken over by the Polish State Railways have their Ty2 numbers shown in italics. This data has been copied from what we believe to be the plan for the circa-1947 renumbering. It should be noted that about 35% of these numbers may be incorrect. Non-italic numbers for Ty2-1 to Ty2-1206 are derived from chassis works-numbers, and can be assumed to give the true identity of each locomotive.

Principal sources of information are listed in the Bibliography which can be found at the end of the book.

Abbreviations

Names of Railways

Ba	Baltic Railway (SZD) 1953-1956. Pribaltic Railway (SZD) 1963-1991.
BDZ	Bulgarian State Railways.
Be	(with date): Byelorussian Railway (SZD).
BL	Brest-Litovsk Railway (SZD) 1944-1951.
CFL	Luxembourg Railways.
CFR	Rumanian State Railways.
CSD	Czechoslovak State Railways.
DB	German Federal Railway 1949-1993.
Do	Donetsk Railway (SZD).
DR	German State Railway (DDR) 1949-1993.
DRB	German State Railway 1920-1945.
DR(o)	German railways in Soviet zone 1945-1949.
DR(w)	German railways in British/US zones 1945-1949.
Es	(with date): Estonian Railway (SZD), 1944-1953, 1956-1963.
FS	Italian State Railways.
GKB	Graz–Köflacher railway (Austria).
Go	Gorki Railway (SZD).
GySEV	Györ–Sopron–Ebenfurti railway.
HDZ	Croatian State Railways.
JDZ	Jugoslav State Railways (1931-c1953).
JZ	Jugoslav Railways (c1953-).
Ki	Kishinev Railway 1944-1953 (SZD).
Kln	Kalinin Railway (SZD).
Kow	Kovel Railway (SZD).
La	(with date): Latvian Railway (SZD), 1944-1953, 1956-1963.
Len	Leningrad Railway 1947-1953 (SZD).
LG	Lithuanian Railways since 1991, (refers to locos in stock 3-1-1993).
Li	(with date): Lithuanian Railway (SZD), 1944-1953, 1956-1963.
Lw	Lvov Railway (SZD).
MAV	Hungarian State Railways.
MK	Moscow–Kiev Railway (SZD).
Mo	Moldavian Railway (SZD) 1956-1961,1979-1991.
MPS	Ministry of Ways of Communication (USSR).
Ms	Moscow Railway (SZD).
NK	North Caucasus Railway (SZD).
NSB	Norwegian State Railways.
ÖBB	Austrian Federal Railways (1947-).
Od	Odessa Railway (SZD), 1944-1953, 1956-1961, 1979-.
OK	Odessa–Kishinev Railway (SZD), 1953-1956, 1963-1979.
Okt	October Railway (SZD).
Om	Omsk Railway (SZD).
ÖStB	Austrian State Railways (1945-1947).
PKP	Polish State Railways.
S	Southern Railway (SZD).
SAAR	Railways of Saarland (1947-1957).
SDZ	Serbian State Railways.
SNCB	Belgian National Railways.
SNCF	French National Railways.
St	Stalin Railway (SZD).
Sw	Sverdlovsk Railway (SZD).
SW	South Western Railway (SZD).
SWDE	Südwestdeutsche Eisenbahnen, ie Railways of the French zone of Germany 1947-1952.
SZD	Soviet Railways.
TCDD	Turkish State Railways.
W	Western Railway (SZD).
Wnz	Vinnitsa Railway (SZD).

Other Abbreviations

Be	(without date): inherited by Belarus from USSR in 1991.
Bw	Bahnbetriebswerk / Locomotive Depot.
BZA	Bundesbahnzentralamt: central research establishment of DB.
CCCPc	Property of Soviet forces in Czechoslovakia: - subsequent fate unknown.
CCCPd	Property of Soviet forces in Germany: - subsequent fate unknown.
CCCPm	Property of Soviet forces in Hungary: taken away mostly in 1948 and 1949.
CCCPö	Property of Soviet forces in Austria: - subsequent fate unknown.
CN	Scrapyard north of Nysa (Poland).
CSSR	Czechoslovak Socialist Republic.
DDR	German Democratic Republic.
Dsp	Dampfspender: "Stationary Boiler".
DWM	Deutsche Waffen- und Munitionsfabriken, Cegielski's works in Poznan 1939-1945.
Es	(no date): inherited by Estonia from USSR in 1991.
Explosion	Locomotive's boiler exploded.
Export	Loco exported from USSR circa 1960.
i	Industry.
IF	Ivano-Frankovsk Works (Ukraine).
K	boiler.
KrV	Written-off due to war damage.
KSch	War damage.
Kx	locomotive without boiler.
La	(no date): inherited by Latvia from USSR in 1991.
Li	(no date): inherited by Lithuania from USSR in 1991.
M	Museum loco (ie preserved).
MBA	Maschinenbau und Bahnbedarf AG, Berlin (Orenstein & Koppel).
NW	Novocherkassk Works (USSR).
ÖGEG	Österreichische Gesellschaft für Eisenbahngeschichte (Linz).
Pi	Polish industry.
RAW	Reichsbahn-Ausbesserungswerk / Main repair workshops of DR/DRB.
s	Scrap. Loco already cut-up.
Uk	Loco in the Ukraine (Lvov Rly), summer 1991.
USA	United States of America.
USSR	Soviet Union.
USSR (i)	Soviet industry, or other non-SZD service.
v	sold (eg v 76 means sold in 1976). See supplementary notes, pages 105-106.
WLF	Wiener Lokomotivfabrik AG, Floridsdorf.
Wrk	Workshops.
x	Locomotive scrapped.
z	Withdrawn from service.
+	Permanently withdrawn from service.
§	Snowplough (wheels and chassis ex 52).
#	See supplementary notes, pages 105-106.

52 001

BORSIG

9/42

---

52 001 DB (Göttingen Vbf 51) + 18.10.54

---

52 002 - 52 075

HENSCHEL

9/42 - 12/42

---

52 002 PKP *Ty2-15* *+ 31.03.80*
003 DB (Minden 50) + 18.10.54
004 USSR (Lw 48, Od 48, Lw 49) i 51
005 PKP *Ty2-16*
006 DR 52 1006 (Kamenz 70) + 16.09.77
007 CCCPd
008 JDZ 33-041
009 PKP *Ty2-17* *+ 29.05.73*
52 010 JDZ 33-042
011 USSR (Kow 44, beschädigt)
012 DB + 18.10.54
013 PKP
014 PKP *Ty2-19* *+ 21.01.72*
015 DR + 13.01.69
016 PKP *Ty2-20* *+ 01.07.85*
017 PKP *Ty2-21* *+ 21.07.72*
018 USSR (Kln 51, Okt 61) + 01.06.71
019 USSR (BL 47, Be 53, Okt 59) + 01.04.73
52 020 USSR (Kow 46, Lw 53) + 01.08.68
021 USSR, La 1042.001 (Jelgava 93) M
022 DB + 18.10.54
023 USSR, La 1042.002 (Riga 92) + 87 x 93
024 USSR (Kow 46, Lw 53, Li 58) + 01.05.72
025 PKP *Ty2-22* *+ 05.03.92*
026 PKP *Ty2-23* *+ 28.12.89*
027 PKP *Ty2-24* *+ 21.06.79*
028 PKP *Ty2-25* *+ 21.03.74*
029 DB + 18.10.54
52 030 PKP *Ty2-26* *+ 11.09.87*
031 USSR 1042.003 (Zheleznodorozhnyj 93)
032 DR 52 1032 (Zwickau 70) + 30.07.75
033 PKP *Ty2-27* *+ 20.02.88*
034 USSR, (63) CSD 555.0297 555.3297 + 68
035 USSR (SW 47, ohne Kessel)
036 USSR, La (Ieriki 92) 1042.612 M
037 USSR (Od 44) i 45
038 DR 52 1038 (Dresden 70) + 20.02.75
039 USSR (Okt 51) + 01.02.73
52 040 USSR (SW 48)
041 JDZ 33-043
042 DR 52 1042 (Kamenz 70) v 01.04.76
043 USSR (SW 47, Ki 48, Lw 49, La 61) + 01.11.69
044
045 DB + 18.10.54
046
047 PKP *Ty2-28* *+ 08.04.74*
048 DB + 18.10.54
049 DB + 18.10.54
52 050 DB + 18.10.54
051 DB + 18.10.54
052 USSR (Kln 53, Be 57) , (63) Pi
053 PKP
054 DR 52 1054 (Hoyerswerda 70) v 10.09.76
055 PKP *Ty2-30* *+ 08.09.73*
056 PKP
057 CCCPm
058 USSR (Li 48, Lw 49, Kln 51, Okt 55) + 01.02.72
059 PKP *Ty2-31* *+ 20.01.73*

52 060 PKP *Ty2-32* *+ 14.12.74*
061 USSR 1042.005 (Mamonowo 93)
062 CCCPm
063 DR(o) 09.45 → 52 5372" DR (Salzwedel 70)+ 20.07.78
52 063" (ex 52 5372) → USSR, (64) JZ 33-296
064 DB + 18.10.54
065
066 ÖStB, ÖBB 152.066 (Linz 63) + 22.11.72
067 PKP *Ty2-33* *+ 20.11.85*
068 USSR 1042.006 (Tschernjachowsk 85) + 10.85
069 USSR (St 48), (51) → TE 8036 #
52 070 PKP *Ty2-34* *+ 18.11.86*
071 USSR (Od 44, Ki 47, OK 53, Od 56, La 58) + 81
072 DR + 04.08.69
073 USSR, (63) CSD 555.0251 555.3251 + 71
074 USSR, Be 1042.007 (Wysokoje 93)
075 DR 52 1075 (Zwickau 70) + 19.06.75

---

52 076 - 52 085

JUNG

1/43 - 2/43

---

52 076 PKP *Ty2-89* *+ 17.07.87*
077 CCCPm → CFR 150.1109 ?
078 PKP *Ty2-35* *+ 25.03.88*
079 DR 52 1079 (Wittenberg 70) + 14.04.76
52 080 USSR, Uk 1042.008
081 USSR (BL 44, W 46, Li 47, Ba 53, Es 56) + 71
082 USSR (Saki 46) → Iwano-Frankowsk Wrk 51 #
083
084 DR 52 1084 (Kamenz 70) + 24.10.75
085 PKP *Ty2-90* *+ 12.10.82*

---

52 086 - 52 123

KRAUSS-MAFFEI

2/43 - 3/43

---

52 086 SWDE, DB (Freudenstadt 50) + 18.10.54
087 USSR 1042.009 (Bagrationowsk 93)
088 ÖStB, ÖBB 152.088 (Amstetten 63) + 22.11.72
089 USSR, (64) JZ 33-286
52 090 DB + 18.10.54
091 ÖStB + 18.03.46
092 DR 52 1092 (Riesa 70) + 06.02.78
093 USSR 1042.010 (Zheleznodorozhnyj 93)
094 DB + 18.10.54
095 DB + 18.10.54
096 DB + 18.10.54
097 DB + 18.10.54
098 USSR 1042.011 (Vilnius 85) + 85
099 PKP *Ty2-91* *+ 07.06.86*
52 100 JDZ 33-044 M
101 PKP *Ty2-92* *+ 18.12.85*
102 USSR (Li 45, La 47, Ba 53, Es 56) i 64
103 USSR,Be 1042.012 (Wysokoje 93)
104 PKP *Ty2-93* *+ 15.05.90*
105 DR 52 1105 (Glauchau 70) + 03.08.73
106 USSR 1042.013 (Bagrationowsk 93)
107 USSR (Len 52, Okt 53, Es 56) + 01.05.74
108 JDZ 33-045
109 USSR (Len 51, Okt 53) i 72
52 110 USSR (Wnz 44) PKP *Ty2-94*
111 PKP *Ty2-95* *+ 04.07.79*
112 DB + 18.10.54
113 PKP *Ty2-96* *+ 15.03.84*
114 USSR 1042.014 (Zheleznodorozhnyj 93)

52 115 DR 52 1115 (Dresden 70) + 16.05.77
116 USSR (SW 44)
117 USSR (Kiew 46) + 51 ?
118 PKP *Ty2-97* *+ 15.06.89*
119 DB + 18.10.54
52 120 USSR, LG (Radviliskis 92)
121 PKP *Ty2-98* *+ 15.08.84*
122 PKP *Ty2-99* *+ 18.11.86*
123 SNCF 150 Y 123

---

52 124 - 52 143

HENSCHEL

1/48 - 11/49

---

52 124 DB (Minden 51) + 18.10 54
125 DB (Minden 51) + 18.10.54
126 DB (Minden 51) + 18.10.54
127 DB (Minden 51) + 18.10.54
128 DB (Minden 51) + 18.10.54
129 DB (Minden 51) (57) Kessel → 50 3165
52 130 DB (Minden 51) (57) Kessel → 50 3166
131 DB (Minden 51) (57) Kessel → 50 3167
132 DB (Minden 51) (57) Kessel → 50 3168
133 DB (Minden 51) (57) Kessel → 50 3169
134 DB (Minden 51) (56) Kessel → 50 3170
135 DB (Minden 51) (56) Kessel → 50 3171
136 DB (Minden 51, Wedau 55) + 11.12.61
137 DB (Minden 51, Wedau 55) + 04.12.61
138 DB (Minden 51, Wedau 55) + 28.05.63
139 DB (Minden 51, Löhne 55) + 04.12.61
52 140 DB (Minden 51, Wedau 55) + 08.10.62
141 DB (Minden 51, Wedau 55) + 08.10.62
142 DB (Minden 51, Wedau 55) + 04.12.61
143 DB (Minden 51, Wedau 55) + 18.05.63

---

52 144 - 52 240

SCHWARTZKOPFF

11/42 - 2/43

---

52 144
145 USSR (BL 44, Be 46, Kohlenindustrie 57) + 73
146 USSR (Be 58), Uk (Wjerbka 89)
147 (CSD 555.0120) ?
148
149 DR 52 1149 (Schöneweide 70) + 03.04.78
52 150 CSD 555.0115 + 70
151 DR (56) BDZ 15.91 (Asenowo 92)
152 CCCPö Tender → CFR
153 USSR 1042.016 (Krasnowka 93)
154 USSR (Li 45, BL 48, Be 53, i 64 Ural Gebiet)
155 USSR 1042.017 (Vilnius 85) + 85
156 DR (56) BDZ 15.86
157
158 USSR (64) BDZ 15.211
159 PKP *Ty2-36* *+ 05.03.76*
52 160 PKP *Ty2-37* *+ 21.07.87*
161 USSR 1042.018 (Kaunas 85)
162 PKP *Ty2-38* *+ 02.04.91*
163 PKP *Ty2-39*
164 DB + 18.04.47

52 165 PKP *Ty2-40* *+ 26.03.84*
166 PKP *Ty2-41* *+ 24.11.72*
167 DB + 18.10.54
168 SWDE, DB (Mainz 50) + 18.10.54
169 USSR (BL 47, Lw 49, IF 68)
52 170 USSR (Kow 46, Lw 53, Li 58) i 63
171 PKP *Ty2-42* *+ 19.10.74*
172 DR, (56) BDZ 15.97 (Asenowo 92)
173 DR, (56) BDZ 15.98 (Kalojanowetz 92)
174 USSR (Okt 51, i 67 Waldindustrie Kirow Gebiet) + 71
175 PKP *Ty2-43* *+ 19.11.91*
176 USSR (Kow 45)
177 DR + 02.01.69
178 DR 52 1178 (Falkenberg 70) + 07.08.79
179 USSR (Kow 46, Wnz 52, SW 53) i 56
52 180 PKP *Ty2-44* *+ 06.10.77*
181 USSR (Lw 51) + 51 ?
182 DR + 03.06.54
183 DR, (56) BDZ 15.89 (Sindel 90)
184 USSR (Li 47, Ki 48, OK 53, Mo 56, Li 60) + 01.10.72
185 USSR, (64) JZ 33-301
186 PKP *Ty2-45* *+ 07.01.75*
187 PKP *Ty2-46* *+ 17.03.77*
188 USSR,Uk
189 ÖStB, ÖBB 152.189 (Knittelfeld 63) + 01.08.70
52 190 PKP *Ty2-47* *+ 15.12.87*
191 PKP *Ty2-48* *+ 13.06.89*
192 DR 52 1192 (Engelsdorf 70) + 12.07.76
193 SWDE, DB (Friedrichshafen 50) + 18.10.54
194 ÖStB, ÖBB 152.194 (Amstetten 63) + 20.10.72
195 DB + 18.10.54
196 CCCPm (51) → CFR 150.1105 (Sibiu 95) M
197 USSR (Kow 45, Lw 53, Li 57, i 74 Kochanowka)
198 DR + 10.06.69
199 PKP *Ty2-49* *+ 05.03.71*
52 200 PKP *Ty2-50*
201 PKP *Ty2-100* *+ 19.01.72*
202 DB + 18.10.54
203 PKP *Ty2-51* *+ 01.10.80*
204 ÖStB, ÖBB 152.204 (Amstetten 63) + 06.05.67
205 USSR, Be 1042.020 (Luninez 93)
206 JDZ 33-046
207 USSR (Apostolowo 46)
208 USSR (Lw 50) 63 → Ausbessungswerk Lwow
209 USSR 1042.021 (Rezekne 90)
52 210 PKP *Ty2-101* *+ 10.06.68*
211 PKP *Ty2-52* *+ 28.12.89*
212 PKP *Ty2-102* *+ 13.05.75*
213 USSR (Ki 44, NW46, Be46, MK 53, Be55, Li 57) i 58
214 USSR 1042.022 (Riga 85)
215 USSR (St 44, Okt 45, Len 47, La 48, Ba 53) + 75
216 DR 52 1216 (Eilenburg 70) v 27.01.76
217 NSB 217 + 04.70
218 USSR 1042.023 (Tschernjachowsk 85)
219 USSR + 4.88
52 220 PKP *Ty2-103* *+ 08.07.85*
221 ÖStB, ÖBB 152.221, (63) 52.221 + 04.77 M
222
223 DB + 18.10.54
52 223" USSR (Lw 46, Ki 48, OK 53, Mo 56, Lw 59) + 73
224 USSR (Li 47) i 51
225 USSR (Li 47, La 48, Ba 53, La 56, i 64) + 73
226 ÖStB + 27.12.46
227 JDZ 33-047
228 USSR (Li 47, Ba 53, La 56, Kln 58, Okt 61, Lw 61)
(62) DR 52 1228 (Wustermark 70) + 21.02.77
229 USSR + 88
52 230 USSR 1042.026 (Zheleznodorozhnyj 93)
231 DB + 18.10.54
232 ÖStB, ÖBB 152.232 (Wien Ost 63) + 18.01.69
233 ÖStB, ÖBB 152.233 (Amstetten 63) (72) → GKB + 72
234 DR 52 1234 (Pasewalk 70) v 17.12.76
235 JDZ 33-048
236 USSR (62) DR 52 1236 (Brandenburg 70) + 19.06.75
237 ÖStB, ÖBB 152.237 (Wiener Neustadt 63) + 20.10.72
238 USSR (W 51, Kln 53, Be 57, i 60 Ölindustrie)
239 USSR 1042.027 (Valga 89) + 88
52 240 SWDE, DB (Friedrichshafen 50) + 18.10.54

52 241 - 52 343

Wiener Lokomotivfabrik

11/42 - 2/43

---

52 241 USSR (Wnz 44-47)
242 DB + 21.04.52
243 PKP *Ty2-53* *+ 27.04.91*
244 SNCF 150 Y 244
245 PKP *Ty2-54* *+ 09.08.90*
246 USSR (Ungeny 78)
247 DR (56) BDZ 15.96 (Kalojanowetz 92)
248 DR (56) BDZ 15.101 (Bodrowo 92)
249
52 250 USSR,Uk
251 PKP *Ty2-55* *+ 28.12.89*
252 NSB 252 + 24.09.63
253 CSD 555.0112
254 USSR (Kow 46, Lw 53, La 57) + 01.10.71
255 USSR, (63) CSD 555.0296 555.3296 + 67
256 USSR (Wnz 44, Lw 47, Ki 48) i 51
257 USSR Uk 1042.029 (Wjerbka 89)
258 PKP *Ty2-56* *+ 10.07.85*
259 PKP *Ty2-57* *+ 15.01.88*
52 260 USSR Uk
261 DB + 18.10.54
262 DB + 20.09.48
263 PKP *Ty2-58*
264 DB + 21.04.52
265 USSR (64) JZ 33-266
266 USSR 1042.031 (Zheleznodorozhnyj 93)
267 PKP *Ty2-59* *+ 28.12.89*
268 PKP *Ty2-60* *+ 17.07.84*
269 USSR (SW 53, i 57 Kamenez-P Elektrizitätswerk)
52 270 DB + 18.10.54
271 USSR (Lw 44 - 48)
272 PKP *Ty2-61* *+ 14.04.82*
273 DB + 18.10.54
274 PKP *Ty2-62* *+ 08.07.86*
275 DR 52 1175 (Jüterbog 70) + 24.10.75
276 PKP *Ty2-63* *+ 14.01.87*
277 CSD 555.0166 555.3166 + 71
278 USSR (Len 51, Okt 53) + 01.03.72
279 PKP *Ty2-64* *+ 26.04.91*
52 280 USSR (W 51, Kln 53, Okt 61) + 01.01.81
281 USSR, (64) BDZ 15.262
282
283 USSR (Li 50, Ba 53, Li 56, i 74 Kaliningrad Seehafen)
284 ÖStB, ÖBB 152.284, (63) 52.284 + 10.10.69
285 SWDE, DB (Betzdorf 50) + 18.10.54
286 USSR, (63) MAV 520,5001 + 12.71
287 DB + 18.10.54
288 ÖStB, ÖBB 152.288 (Amstetten 63) + 20.10.72
289 DB (Osnabrück Gbf. 50) + 18.10.54
52 290 USSR (Ki 48, OK 53, Mo 56, La 57, Li 59) + 01.04.75
291 SWDE, DB (Worms 50) + 18.10.54
292 JDZ 33-180
293 USSR (Od 49, OK 53, Ba 55, Li 56) + 01.02.74
294 DB + 18.10.54
295 DB + 18.10.54
296 USSR (Kow 45), PKP *Ty2-65* *+ 87*
297 PKP *Ty2-66* *+ 07.03.87*
298 USSR 1042.032 (Tschernjachowsk 85)
299 DB + 18.10.54
52 300 DB (Coesfeld 50) + 18.10.54
301 ÖStB, ÖBB 152.301 (Wiener Neustadt 63) (73) →GKB + 76
302 DR 52 1302 (Görlitz 70) + 28.10.77
303 USSR (Od 44, → Buntmetallindustrie: Kupferkombinat Rewda, '63 → Ural Aluminiumwerk) + 09.66
304 DR 52 1304 (Dresden 70) + 17.02.76
305 USSR (Ba 53, La 56, i 76 Zuckerfabrik Jekabpils)
306 USSR (Okt 51, Es 57, i 59 Elektrizitätswerk Narva)
307 USSR 1042.033 (Zheleznodorozhnyj 93)
308 CSD 555.0160, 555.3160 + 69
309 ÖStB, ÖBB 152.309, (68) GKB + 75
52 310 PKP *Ty2-67*
311 USSR (St 44-48, Lw 49, Li 59) + 01.08.72
312 USSR (Od 45) + 55
313 USSR 1042.034 (Radviliskis 85) → LG
314 PKP *Ty2-68* *+ 12.05.81*
315 USSR 1042.035 (Krasnowka 93)
316 CSD 555.0164, 555.3164 + 70
317 USSR (Be 47, Lw 62), (64) JZ 33-302
318 USSR 1042.036 (Jelgava 85) + 85
319 USSR (S 44-46)
52 320 USSR, (66) JZ 33-330
321 USSR (OK 53, Od 56, Es 57) + 01.07.75
322 USSR, Be 1042.037 (Neman 93)
323 DR 52 1323 (Pasewalk 70) + 16.09.77
324 NSB 324 + 24.09.63
325 DR 52 1325 (Eberswalde 70) + 24.03.77
326 USSR (Od 52, OK 53, i 56 Kohlenindustrie)
327 PKP
328 PKP *Ty2-105* *+ 17.07.87*
329 PKP *Ty2-106* *+ 17.03.90*
52 330 PKP *Ty2-107* *+ 13.03.87*
331 CSD 555.0163, 555.3163 + 72
332 PKP *Ty2-108* *+ 15.01.88*
333 DB + 18.10.54
334 PKP *Ty2-109* *+ 14.12.81*
335 USSR, Be 1042.038 (Wysokoje 93) M
336 DR 52 1336 (Wustermark 70) + 15.03.73
337 USSR (SW 50, S 53, i 56 Kohlenindustrie)
338 PKP *Ty2-110* *+ 10.01.86*
339 CSD 555.0167 + 69
52 340 USSR 1042.039 (Vilnius 85) + 88
341 CSD 555.0162 555.3162 + 70
342 DR 52 1342 (Bautzen 70) + 16.09.77
343 USSR (Lw 45)

---

52 344 - 52 349

HENSCHEL

12/42

---

52 344 DR 52 1344 (Kamenz 70) + 27.10.71
345 USSR (Be 48, W 51, Kln 53, i 56 Kohlenindustrie)
346 DR 52 8158 (Brandenburg 70) + 15.08.88
347 USSR (Okt 53, i 67, Go 69) + 01.03.70
348 DB + 18.10.54
349 PKP *Ty2-69* *+ 24.11.88*

---

52 350 - 52 525

BORSIG

52 350 - 523 4/43 - 9/43
52 524 - 525 1/44

---

52 350 DB (Göttingen Vbf. 50) + 18.10.54
351 CCCPd
352 DR(o) + 46
353 USSR (Li 48)
354 ÖStB, ÖBB + 20.11.54
355 PKP *Ty2-111*
356 DR 52 1356 (Dresden 70) + 24.10.75
357 ÖStB, ÖBB + 53, § 985.203
358 DR 52 8043 (Elsterwerda 70) Dsp 88, M
359 USSR (Od 44, ohne Kessel)
52 360 DR 52 1360 (Berlin Schöneweide 70) z 10.87 M
361 USSR, (62) DR 52 1361 (Elsterwerda 70) + 14.05.79
362 DR 52 8126 (Brandenburg 70) z 08.88 M
363 USSR (Be 46, Kln 54, Okt 55) i 73
364 TCDD 56511 (Tavsanli 81, Usak 95)

52 365 TCDD 56512 (Tavsanli 81, Usak 95)
366 TCDD 56513 (Usak 81, Dinar 90)
367 TCDD 56514 (Afyon 81, Alasehir 95)
368 TCDD 56515 (Erzurum 75)
369 USSR (Od 44) i 45
52 370 PKP *Ty2-112* *+ 10.04.92*
371 DR 52 1371 (Falkenberg 70) + 17.02.76
372 PKP *Ty2-113*
373 DR 52 8081 (Lübbenau 70) + 13.07.83
374 DR 52 1374 (Dresden 70) + 27.10.71
375 DR 52 8076 (Altenburg 70) z 08.87
376 PKP *Ty2-114* *+ 19.08.85*
377 PKP Ty2-115 (Zakopane 91) + 08.02.91
378 DR 52 1378 (Neustrelitz 70) + 15.03.73
379 DR 52 1379 (Berlin Schöneweide 70) + 20.05.81
52 380 USSR (Be 58), Uk
381 PKP Ty2-116 (Zakopane 91) + 15.12.92
382 DR 52 1382 (Engelsdorf 70) + 02.03.78
383 USSR (SW 51, S 53) i 56
384 ÖStB, ÖBB + 20.11.54
385 PKP *Ty2-117* *+ 15.11.87*
386
387 PKP *Ty2-118* *+ 03.03.73*
388 PKP *Ty2-119* *+ 02.12.90*
389 USSR
52 390 PKP *Ty2-120* *+ 02.06.87*
391 DB + 18.10.54
392 SWDE, DB (Friedrichshafen 50) + 18.10.54
393 JDZ 33-181
394 PKP *Ty2-121*
395 DR 52 1395 (Wittenberg 70) + 20.07.78
396 PKP *Ty2-122* *+ 10.02.87*
397 DR 52 1397 (Cottbus 70) + 19.04.76
398 USSR, (64) JZ 33-316
399 PKP *Ty2-123* *+ 30.04.87*
52 400 DB + 18.10.54
401 ÖStB, ÖBB + 25.06.56, § 985.204
402 USSR, (63) PKP
403 PKP *Ty2-124*
404 PKP
405 USSR (La 47, Ba 53, La 56) + 01.04.75
406 PKP *Ty2-126* *+ 25.09.92*
407 DR 52 1407 + 27.10.69
408 DB + 18.10.54
409 DR 52 1409 (Templin 70) v 23.03.77
52 410 DB + 18.10.54
411 USSR (Li 47, La 50, Ba 53, La 56) + 01.07.75
412
413 PKP *Ty2-127* *+ 19.12.85*
414
415 DR 52 9415 (Senftenberg 70) + 01.07.77
416 USSR (Lokostroj 1.1.46, ohne Kessel)
417 JDZ 33-049
418 USSR (Li 46, Ki 48, OK 53, Od 56, Es 57, i 59 Narva)
419 PKP *Ty2-128* *+ 13.04.67*
52 420 DB + 18.10.54
421 USSR (Li 45, La 48, Ba 53) + 3.56
422 USSR (Lokostroj 1.1.46; Rahmen → TE-5694) M
423 USSR (Be 58), Uk (Wjerbka 89)
424 USSR, Be 1042 043 (Brusgi 93)
425
426 PKP *Ty2-129* *+ 23.02.90*
427 DB + 18.10.54
428 PKP Ty2-871 (Warszawa 91)
429 USSR (Od 44) i 45
52 430 DB + 18.10.54
431 DB + 18.10.54
432 DB + 18.10.54
433 ÖStB, ÖBB + 28.10.58 § 985.207
434 USSR (Od 48, OK 53, Od 56, i 56 Kohlenindustrie)
Rahmen →TE-8031
435 USSR (Ungeny 78, Ungeny 94)
436 DR 52 1536 (Berlin Schöneweide 70) + 26.09.69
437 ÖStB, ÖBB + 20.11.54
438 DR 52 1538 (Altenburg 70) + 25.07.86
439 DB + 18.10.54

52 440
441 USSR, (63) MAV 520,001 + 8.65
442 DB + 11.01.52
52 442" USSR (Ki 49, Wnz 51, SW 53, Li 56) i 64
443 PKP *Ty2-132* *+ 20.02.74*
444 DR 52 1444 (Glauchau 70) v 23.04.79
445 PKP *Ty2-133* *+ 02.12.90*
446 PKP *Ty2-134* *+ 15.09.89*
447 USSR (BL 46)
448 PKP *Ty2-135* *+ 15.05.90*
449 DB + 18.10.54
52 450 USSR (Li 47, Ba 53, Es 53) + 01.10.71
451 PKP *Ty2-136* *+ 30.04.87*
452 PKP
453 PKP *Ty2-138* *+ 07.05.74*
454 DR 52 1454 (Pasewalk 70) v 20.12.77
455 DB + 18.10.54
456 USSR TE-456 (62) PKP Ty2-1290
TE-456" (Radviliskis 85) 1042 045
457 USSR,Uk (Rawa Russkaja 92)
458 FS (48) → DB + 18.10.54
459 PKP *Ty2-139* *+ 23.02.90*
52 460 DB, (51) JDZ 33-240 M
461 PKP Ty2-642 (Leszno 93)
462 PKP Ty2-847 (Zakopane 91) + 05.07.91
463 USSR, (63) PKP Ty2-1373 (Wolsztyn 83) + 11.06.87
464 USSR, (63) MAV 520,075 (76-79) GySEV M
465
466 USSR, (63) PKP
467 DR 52 8200 (Templin 70) M
468 DR 52 1468 (Cottbus 70) v 10.09.76
469 USSR (SW 53) i 56
52 470 DR 52 8140 (Brandenburg 70) + 09.08.88
471 PKP *Ty2-142* *+ 26.07.89*
472 DR 52 8182 (Brandenburg 70), Dsp 02.90
473 JDZ 33-050
474 DR 52 8135 (Brandenburg 70) M
475 JDZ 33-051
476 PKP *Ty2-143* *+ 14.06.89*
477 DB + 14.02.51
478 ÖStB, ÖBB (St.Pölten 63) + 20.10.75 § 985.224
479 USSR 1042.047 (Tapa 85) + 87
52 480 PKP *Ty2-144* *+ 07.02.91*
481 PKP *Ty2-145* *+ 10.07.85*
482 USSR (Od 44, Ki 48, OK 53, Ba 54, Es 56) i 58
483 PKP *Ty2-146* *+ 28.11.80*
484 PKP *Ty2-147* *+ 12.09.89*
485 USSR (Wnz 44, Od 48, OK 53, Od 56) i 56
486
487 PKP *Ty2-148* *+ 22.11.83*
488 DB + 18.10.54
489
52 490 ÖStB, ÖBB + 20.11.54
491 PKP *Ty2-149* *+ 23.01.71*
492 USSR (Kln 51, Okt 61) + 01.02.71
493 DR 52 8192 (Eberswalde 70) x 01.90
494 USSR, (64) JZ 33-267
495 USSR (Be 58), Uk
496 PKP *Ty2-150* *+ 09.06.76*
497 USSR (Be 51, Kln 54) i 56
498 USSR (Kow 46, Wnz 52, SW53) i 57
499 ÖStB, ÖBB + 20.11.54
52 500 USSR (Be 58, Ungeny 78)
501 USSR (Wnz 44)
502 DR 52 1502 (Halle-G 70) + 02.06.76
503 PKP *Ty2-151* *+ 15.05.90*
504 USSR (BL 47, Be 53, Be 58) + 68
505 USSR (Riga Skirotava 46, KrV) + 08.46
506 PKP *Ty2-152* *+ 22.02.90*
507 USSR 1042.050 (Tapa 85)
508 BDZ 15.01 (Kalojanowetz 92)
509 USSR (Be 50)
(62) DR 52 1509 (Brandenburg 70) + 14.04.76
52 510 USSR, (63) PKP Ty2-1394 (Walbrzych 83) + 87
511 USSR (Od 47, OK 53, i 56 Kohlenindustrie)
512 USSR ?
513 USSR (SW 48) i 56
514 PKP *Ty2-153* *+ 01.11.81*

52 515	PKP	*Ty2-154*	*+ 28.11.80*
516	ÖStB, ÖBB		+ 20.11.54
517	USSR	(Ungeny 78)	
518	JDZ	33-052	
519	USSR	1042.052 (Radviliskis 85)	
52 520			
521	USSR	(Be 50, Be 58, SW 59)	i 60
522	USSR	(Li 45)	
523	PKP	*Ty2-155*	*+ 17.03.90*
524	USSR	(Od 48, OK 53, Od 56)	i 57
525	USSR	1042.053 (Vilnius 85)	

---

52 526 - 52 604

SCHWARTZKOPFF

12/43 - 5/44

---

52 526	(CSD	555.068) ?	
527	USSR,	(62) CSD 555.0217 555.3217	+ 70
528	USSR, Uk		
529	USSR,	(62) PKP Ty2-1236	+ 20.07.77
52 530			
531	PKP		
532	USSR,	(62) DR 52 8160 (Brandenburg 70)	M
533	DR	52 9533 (Senftenberg 70)	+ 14.11.79
534	USSR	(SW 48)	i 56
535	PKP	*Ty2-157*	*+ 02.06.87*
536	USSR	(SW 48, i 56 Buntmetallindustrie)	
537	PKP	*Ty2-158*	*+ 12.09.88*
538	DB		+ 18.10.54
539	DR	52 8033 (Halle G 70)	+ 15.08.88
52 540			
541	DR		+ 13.01.69
542	DB		+ 11.07.51
543	PKP	*Ty2-159*	*+ 14.06.85*
544	PKP		
545	DR	52 1545 (Zwickau 70)	+ 19.12.74
546			
547	DR	52 8148 (Hoyerswerda 70)	M
548	PKP	*Ty2-161*	*+ 31.01.88*
549	USSR	(Be46, Li 59), (63) PKP Ty2-1297	+ 24.01.77
52 550	PKP	Ty2-567 (Milkowice 91)	+ 25.11.88
551	PKP	*Ty2-163*	*+ 26.07.91*
552	DR		+ 25.05.54
553	USSR	(La 45-48)	
554	PKP	*Ty2-164*	*+ 08.02.91*
555	CSD	555.0157, 555.0608, 555.0178	+ 69
556	DR	52 8194 (Templin 70)	z 02.88 M
557	USSR	(SW 50 Li 56)	i 57
558	DB		+ 18.10.54
559	DB		+ 18.10.54
52 560	USSR	(Be 50, Kln 54, Li 59), (63) PKP Ty2-1272	
561	USSR	(i) (Lesnaja 63)	
562	PKP	*Ty2-839*	*+ 08.12.78*
563	DR	52 8028 (Neubrandenburg 70)	M
564	DR	52 1564 (Berlin Schöneweide 70)	+ 16.07.80
565	PKP	*Ty2-840*	*+ 15.01.77*
566	PKP	*Ty2-841*	
567	USSR	(i) (Lesnaja 63)	
568	DR	52 1568 (Zwickau 70)	+ 20.02.75
569	PKP	*Ty2-842*	*+ 25.03.87*
52 570	PKP	*Ty2-843*	*+ 10.07.86*
571	USSR	(Li 50, Ba 53, Es 56)	+ 01.08.73
572	DR(o)		+ 04.46
573	DR	52 1573 (Frankfurt/Oder 70)	v 74
574	DB		+ 18.10.54
575	USSR	Be 1042.055 (Wysokoje 93)	
576	DB		+ 18.10.54
577	USSR	(OK 53, Ba 55, Li 56)	+ 01.07.74
578	DR	52 8166 (Stendal 70)	Dsp 04.89
579	ÖStB, ÖBB		+ 20.11.54
52 580	ÖStB, ÖBB	(Mürzzuschlag 63)	+ 29.11.77
581	JDZ	33-053	
582	PKP	*Ty2-844*	*+ 23.02.90*
583	PKP	*Ty2-845*	*+ 20.11.87*
584	DR(o)		+ 14.04.48
585	USSR, Uk	(Wjerbka 89)	
586	PKP	Ty2-815 (Zagorz 88)	
587	PKP	*Ty2-848*	*+ 12.81*
588	DR		+ 29.08.68
589	PKP	*Ty2-849*	*+ 25.03.87*
52 590	PKP	*Ty2-850*	*+ 27.08.74*
591	PKP	*Ty2-851*	*+ 05.02.83*
592	PKP	*Ty2-852*	*+ 03.05.86*
593	USSR	(La 50)	i 51
594			
595	PKP	Ty2-853 (Gniezno 93)	+ 07.10.92
596	PKP	*Ty2-854*	*+ 12.09.88*
597	PKP	*Ty2-855*	
598	PKP	*Ty2-856*	*+ 12.03.86*
599	PKP	*Ty2-70*	*+ 15.09.89*
52 600	DR	52 1600 (Berlin Pankow 69)	v 29.11.71
601	PKP	*Ty2-857*	*+ 15.01.77*
602	USSR	(Be 58), (63) PKP	
603			
604	DR	52 1604 (Brandenburg 70)	+ 06.09.75

---

52 605 - 52 691

SCHICHAU

10/43 - 5/44

---

52 605	USSR	(La 47, Kln 51, Li 58, i 73 Akmene)	
606	USSR	(Preußisch Eylau 47)	+ 48
607	USSR	1042.057 (Krasnowka 93)	
608	SWDE, DB	(Mainz 50)	+ 18.10.54
609	PKP	*Ty2-165*	
52 610	DR	52 1610 (Salzwedel 70)	+ 11.10.76
611	USSR	(Wnz 44)	
612	ÖStB,	ÖBB 152.612	+ 28.10.54
613	USSR	Rahmen → TE-8004	
614	DR	52 8105 (Altenburg 70)	+ 06.03.91
615	PKP		
616	DR	52 1616 (Dresden 70)	+ 14.04.76
617	PKP	*Ty2-167*	*+ 21.04.82*
618	PKP	*Ty2-168*	*+ 08.09.82*
619	DB		+ 18.10.54
52 620	USSR	(i) → Go 69	+ 01.02.70
621	DB		+ 18.10.54
622	USSR	1042.058 (Gusew 92)	M
623	USSR	(Okt 53, Be 58), (63) PKP	
624	USSR	(Okt 53, Li 62), (63) PKP Ty2-1359?	
625	USSR,	(62) CSD 555.0209 555.3209	+ 69
626	DR	52 1726 (Hoyerswerda 70)	+ 12.07.76
627	DR	52 8020 (Elsterwerda 70), Dsp 08.88	M
628	PKP	Ty2-858	+ 15.12.87
629	DR		+ 27.12.71
52 630	DR	52 1630 (Bautzen 70)	+ 08.04.86
631	PKP	*Ty2-859*	*+ 08.02.85*
632	DR	52 8186 (Stendal 70)	v 04.92 M
633	PKP	*Ty2-860*	*+ 10.04.92*
634	USSR	(Li 45-47, Es 47, Li 48, Kaliningrad Wrk 71)	
635			
636	USSR, Uk		
637			
638	USSR	(Li 47-50)	+ 51
639	CCCPd		
52 640	USSR	1042.060 (Tschernjachowsk 85)	
641	PKP	*Ty2-861*	*+ 15.05.90*
642	USSR	(Li 45)	
643	USSR	1042.061 (Kaliningrad 85)	
644	PKP	*Ty2-862*	*+ 01.08.85*

52 645	USSR (La 48, Ba 53, La 56)	+ 01.12.73
646	USSR 1042.062 (Tschernjachowsk 85)	
647	USSR 1042.063	+ 88
648	USSR (Li 48-50)	+ 51
649	USSR, (64) BDZ 15.250	
52 650	USSR, (62) CSD 555.0212 555.3212	+ 68
651	DB	+ 18.10.54
652	BDZ 15.136 oder PKP *Ty2-71* ?	
653	DR 52 9653 (Senftenberg 70)	+ 06.03.79
654	USSR, (64) BDZ 15.162	
655	USSR (Be 58, Ungeny 78)	
656	USSR (Ungeny 78, Ungeny 94)	
657	USSR (Ba 53, La 56, Lw 59)	i 65
658	PKP *Ty2-863*	*+ 05.10.90*
659	PKP *Ty2-864*	*+ 06.02.73*
52 660	NSB 660	+ 04.70
661	DB	+ 18.10.54
662	DR 52 1662 (Cottbus 70)	v 91 M
663	USSR, (64) BDZ 15.182 (Kalojanowetz 92)	
664	DB	+ 18.10.54
665	USSR, (63) MAV 520,5002	+ 08.71
666	PKP *Ty2-865*	*+ 28.09.73*
667	USSR (Li 45-48, SW 48, S 53, i 56 Kohlenindustrie)	
668	USSR	i 52
669		
52 670	USSR (Li 45)	
671	DR 52 8001 (Lübbenau 70)	z 08.88 M
672	USSR, (63) MAV 520,056	+ 07.72
673	PKP *Ty2-866*	*+ 18.08.76*
674	USSR (La 48, Ba 53, La 56)	i 63
675	PKP *Ty2-867*	*+ 06.11.84*
676	USSR, (63) PKP Ty2-1384 (Andrzejowka 91)	+ 91
677	USSR, (64) BDZ 15.263 (Tulowo 90)	
678	PKP *Ty2-868*	
679	USSR, (64) BDZ 15.225 (Kalojanowetz 92)	
52 680	DB	+ 18.10.54
681	PKP	
682		
683	USSR (Mo 56), BDZ 15.164	
684		
685		
686	USSR (Li 45, SW 48)	i 55
687	USSR 1042.066 (Radviliskis 85)	
688	PKP *Ty2-870*	*+ 04.11.89*
689	USSR, (63) PKP Ty2-1371 (Jaworzyna 91)	+ 29.04.91
52 690	USSR (Li 48)	
691	USSR 1042.067 (Tschernjachowsk 85)	+ 86

52 692 - 52 817

HENSCHEL

3/44 - 8/44

52 692	DR (o)	+ 08.47
693	JDZ 33-223	
694	BDZ 15.02	
695	DB (Bayreuth 50)	+ 18.10.54
696	USSR (W 51, Kln 53, Okt 61)	+ 01.07.73
697	BDZ 15.03	
698	PKP Ty2-1124 (Tczew 90)	+ 17.03.91
699	BDZ 15.04 (Asenowo 92)	
52 700	BDZ 15.05 (Bodrowo 92)	
701	KrV (Riesa 46)	+ 46
702	JDZ 33-054	
703	DR 52 1703 (Salzwedel 70)	+ 17.02.76
704	DR 52 8050 (Frankfurt / Oder 70)	x 09.89
705	DR 52 1705	+ 29.09.80
706	USSR (Be 47, SW 48)	i 56
707	DR 52 1707	+ 12.07.76
708	PKP *Ty2-872*	*+ 20.02.88*
709		
52 710	USSR (Be 57, Lw 59), (63) MAV 520,008	+ 12.80
711	DR 52 8093 (Lübbenau 70)	x 12.91
712	DR 52 8037 (Falkenberg 70)	z 05.88 M
713	DR	+ 03.11.69
714	DR 52 1714 (Berlin Schöneweide 69)	+ 27.12.71
715	DR 52 1715 (Salzwedel 70)	+ 24.10.75
716	DR 52 9716 (Senftenberg 70)	+ 14.05.79
717	USSR (Lw 48, Od 52, OK 53, i 56 Kohlenindustrie)	
718	DR 52 8046 (Elsterwerda 70)	z 11.90
719	DR 52 1719 (Neustrelitz 70)	+ 14.07.75
52 720	USSR (Od 56, Mo 58), BDZ 15.171	
721	USSR, Be 1042.068 (Neman 93)	
722	USSR (Shlobin 47), BDZ 15.158	
723	USSR (La 47, Ba 53, La 56, i 74 Zuckerfabrik Liepaja)	
724	USSR (La 47, Ba 53, La 56, i 72 Riga)	
725	FS (48) DB, (51) JDZ	
726	ÖStB KrV	+ 09.02.47
727	ÖStB, ÖBB	+ 28.10.54
728	DB (51) JDZ 33-243	
729	JDZ 33-218	
52 730	JDZ 33-056	
731	JDZ 33-057	
732	DB	+ 18.10.54
733	BDZ 15.06	
734	JDZ 33-224	
735	USSR, (64) BDZ 15.203	
736	USSR (Ungeny 78)	
737	USSR 1042.070 (Bagrationowsk 93)	
738	USSR (Mo 56), BDZ 15.166	
739	ÖStB, ÖBB (Wien Ost 63)	+ 17.02.75
52 740	USSR, (64) BDZ 15.213 (Sindel 90)	
741		
742	ÖStB, ÖBB	+ 27.11.57
743	CSD 555.0132	+ 69
744	USSR, (64) BDZ 15.193 (Asenowo 92)	
745	USSR (Daugavpils-Wrk 50)	i 50
746	DR 52 1746 (Berlin Schöneweide 70)	+ 13.10.80
747	USSR (OK 53, Od 56, Li 58)	+ 01.10.73
748	DB, (51) JDZ 33-237	
749	USSR (Lesnaja 57)	
52 750	PKP Ty2-873 (Jaworzyna 87)	+ 10.11.87
751		
752	DB	+ 18.10.54
753	PKP *Ty2-874*	*+ 15.05.90*
754	DB	+ 18.10.54
755		
756	USSR, (64) JZ 33-287	
757	USSR, BDZ 15.167	
758	USSR, (64) BDZ 15.186	
759	USSR (Li 58 Be 62), (63) PKP	
52 760	DR, (59) BDZ 15 126 ?	
761	PKP *Ty2-875*	*+ 21.02.88*
762	SWDE, DB (Radolfzell 50)	+ 18.10.54
763	PKP *Ty2-876*	*+ 05.09.73*
764	USSR, (64) BDZ	
765	USSR (Lw 49, La 58, Li 59)	
	52 765" ÖStB (Ex SDZ 33-014), ÖBB + 53	§ 985.200
766	DB	+ 18.10.54
767	PKP *Ty2-877*	*+ 02.10.80*
768	USSR (La 58, Be 60), (62) PKP	
769	USSR	+ 88
52 770	DR	+ 30.11.53
771	USSR (Li 47, SW 48, La 57)	+ 01.05.73
772	USSR, (63) PKP Ty2-1303	
773	USSR (Be 48, Sw 49, Od 57, Lw 61)	
	(62) DR 52 1773 (Brandenburg 70)	+ 24.10.75
774	USSR (MK 53, Be 55), (63) PKP	
775	PKP Ty2-878 (Jaworzyna 84)	+ 18.08.89
776	USSR, (63) MAV 520,057 #	+ 08.74
777	USSR 1042.072 (Kaliningrad 93)	M
778	CSD 555.059, 555.3059	+ 72
779	USSR (Bw Pomoschnaja 64)	+ 01.11.78
52 780	CSD 555.0133 ?	
781	USSR (Liepaja 78)	i 84
782	USSR, (64) JZ 33-268	
783	CSD 555.0134 555.3134	+ 67
784	CSD	

52 785	CCCPm KrV	
786	USSR (Ungeny)	
787		
788	JDZ 33-058	
789	JDZ 33-059	
52 790	JDZ 33-060	
791	CCCPö	
792	JDZ 33-061	
793	USSR (Shlobin 47), (64) Jugoslawien 33-504 (Ahmici 96)	
794	PKP *Ty2-880*	*+ 92*
795	USSR (Kln 58, Li 59), (63) PKP	
796	ÖStB KrV	+ 09.02.47
797	ÖStB KrV	+ 25.01.48
798	ÖStB, ÖBB (Selzthal 63)	+ 25.04.73
799	USSR 1042.074 (Tschernjachowsk 85)	
52 800	USSR, (62) CSD 555.0239 555.3239	+ 70
801	ÖStB, ÖBB	+ 28.10.54
802	USSR, Uk	
803	ÖStB, ÖBB (St.Pölten 63)	+ 15.07.70
804	USSR Be 1042.076 (Wysokoje 93)	
805		
806	PKP *Ty2-881*	*+ 14.02.79*
807	USSR (Li 47, Ba 53, La 56, i 73 Riga)	
808	DR 52 1808 (Lichtenberg 70)	+ 12.07.76
809	DR 52 1809 (Wustermark 70)	+ 26.09.69
52 810	USSR (Be 46, Okt 58, Be 60), (63) PKP Ty2-1252	
811	USSR (Kln 51, Okt 55, Li 58), (63) PKP	
812	DB	+ 18.10.54
813	PKP Ty2-882 (Chabowka 88)	+ 07.09.88
814	USSR (Od 48, OK 53, Od 56, i 57 Kohlenindustrie)	
815	USSR 1042.077 (Kaliningrad 93)	
816	USSR 1042.078 (Mamonowo 93)	
817	DB (51) JDZ 33-236 (Tuzla 96)	

52 818 - 52 870

KRENAU

7/44 - 11/44

52 818	PKP *Ty2-883*	*+ 20.09.89*
819	DR 52 8114 (Altenburg 70)	Dsp 90
52 820	USSR (Ba 53, Li 56), (63) PKP Ty2-1389	+ 15.06.92
821	DB	+ 10.08.57
822	USSR (Bulanasch 57)	
823	PKP Ty2-1112 (Bialogard 93)	+ 05.10.90
824	USSR, (62), CSD 555.0250 555.3250	+ 67
825	PKP *Ty2-885*	*+ 16.01.76*
826	USSR, (62) CSD 555.0203 555.3203	+ 68
827	USSR, (62) CSD 555.0205 555.3205	+ 68
828	USSR, (62) CSD 555.0207 555.3207	+ 69
829	USSR, (64) JZ 33-269	
52 830	USSR (SW 48, i 57 Kohlenindustrie)	
831	USSR, Uk (Grodowiza 92)	
832	USSR, (63) CSD 555.0277 555.3277	+ 64
833	CSD 555.085 555.3085	+ 68
834	PKP *Ty2-886*	*+ 06.81*
835	DR 52 1835 (Salzwedel 70)	+ 16.05.77
836	ÖStB, ÖBB (St.Valentin 63)	+ 21.03.75
837	USSR, Be 1042.080 (Luninez 93)	
838	FS (47) DB	+ 07.02.52
839	USSR (Be 58), (62) PKP Ty2-1214	+ 85 → Vietnam
52 840	USSR, Be 1042.081 (Wysokoje 93)	
841	USSR ?	
842	USSR (BL 50, Be 53, Li 59) (62) DR, 52 1842 (Pasewalk 70)	+ 01.02.77
843	JDZ 33-182	
844	ÖStB, ÖBB (Selzthal 63)	+ 25.07.77
52 845	JDZ 33-062	
846	DB	+ 10.08.57
847	JDZ 33-215	
848	USSR, Uk (Rawa Russkaja 92)	
849	PKP *Ty2-887*	*+ 05.07.91*
52 850	USSR (Od 56, La 59), (63) PKP Ty2-1299	+ 14.06.82
851	USSR (Lw 50, Li 58), (63) PKP Ty2-1406	+ 12.09.88
852	PKP *Ty2-888*	*+ 08.02.85*
853	JDZ 33-183	
854	USSR (OK 53 Mo 56), Export	
855	ÖStB, ÖBB (Selzthal 63)	+ 10.78 M
856	USSR 1042.083 (Tapa 85)	
857	USSR, Uk 1042.084	
858	USSR 1042.085 (Zheleznodorozhnyj 93)	
859	ÖStB, ÖBB (Selzthal 63)	+ 12.11.75
52 860	USSR (La 47, Ba 53, La 56, i 56 Kohlenindustrie)	
861	DB	+ 10.08.57
862	ÖStB, ÖBB (Selzthal 63)	+ 04.11.74
863	USSR, Uk 1042.086 (Rawa Russkaja 92)	
864	DB	+ 10.08.57
865	PKP *Ty2-889*	*+ 20.06.81*
866	USSR, (64) BDZ 15.159	
867	USSR, Uk 1042.087 (Rawa Russkaja 92)	
868	USSR (OK 53, Mo 56 Be 59), (63) PKP	
869	PKP *Ty2-890*	*+ 08.06.72*
52 870	USSR, (63) MAV 520,085	+ 06.75

52 871 - 52 874

ESSLINGEN

5/44 - 6/44

52 871	USSR (Li 45, Kow 46, Lw 53, Li 58)	+ 01.03.76
872	DR 52 1872 (Cottbus 70)	+ 09.01.76
873	PKP	
874	USSR (Li 47, Ba 53, Li 56)	+ 01.09.76

52 875 - 52 892

HENSCHEL

1950 - 1951

52 875	DB	(Löhne 51,	Löhne 55)	+ 01.08.61
876	DB	(Löhne 51,	Wedau 55)	+ 29.05.61
877	DB	(Löhne 51,	Wedau 55)	+ 28.03.61
878	DB	(Löhne 51,	Löhne 55)	+ 08.10.62
879	DB	(Löhne 51,	Löhne 55)	+ 04.12.61
52 880	DB	(Löhne 51,	Wedau 55)	+ 04.12.61
881	DB	(Löhne 51,	Löhne 55)	+ 04.12.61
882	DB	(Löhne 51,	Löhne 55)	+ 18.06.62
883	DB	(Löhne 51,	Löhne 55)	+ 01.08.61
884	DB	(Löhne 51,	Wedau 55)	+ 08.10.62
885	DB	(Löhne 51,	Wedau 55)	+ 28.03.61
886	DB	(Löhne 51,	Löhne 55)	+ 04.12.61
887	DB	(Löhne 51,	Löhne 55)	+ 28.05.63
888	DB	(Löhne 51,	Löhne 55)	+ 04.12.61
889	DB	(Löhne 55)		+ 18.06.62
52 890	DB	(Löhne 55)		+ 04.12.61
891	DB	(Bingerbrück 55)		+ 08.10.62
892	DB	(Bingerbrück 55)		+ 04.12.61

52 1098 - 52 1099

MASCHINENFABRIK ESSLINGEN

5/44

---

52 1098 ÖStB, ÖBB (Wien Ost 63) + 17.02.75
1099 ÖStB, ÖBB (Straßhof 63) + 01.08.70

---

52 1100 - 52 1349

DWM

5/43 - 9/44

---

52 1100 NSB 1100 + 24.09.63
1101 NSB 1101 + 04.70
1102 USSR 1042.088 (Bagrationowsk 93)
1103 USSR, Uk 1042.089 (Rawa Russkaja 92)
1104 NSB 1104 + 04.70
1105 DB + 20.11.58
1106 NSB 1106 + 24.09.63
1107 NSB 1107 + 04.70
1108 DR 52 8063 (Leipzig West 70) Dsp 31.05.88
1109 DB + 20.07.59
52 1110 USSR Be 1042.090 (Wysokoje 93)
1111 DR 52 8005 (Eberswalde 70) + 06.09.88
1112 USSR (OK 53, St 56, Od 56, i 56 Kohlenindustrie)
1113 DB + 20.07.59
1114 USSR (Luninez 90) M
1115 USSR (Li 47, Ba 53, Es 56) i 63
1116 PKP *Ty2-169* *+ 08.03.73*
1117 PKP *Ty2-170* *+ 31.01.88*
1118 USSR 1042.092 (Kaliningrad 92)
1119 DB + 10.08.57
52 1120 CSD
1121 DB + 10.08.57
1122 USSR, Uk (Wjerbka 89)
1123 USSR, (62) CSD 555.0248 555.3248 + 67
1124 DB + 10.08.57
1125 PKP Ty2-171 (Jaworzyna 83) + 08.12.92
1126 USSR (Od 44) i 45
1127 CCCPm → CFR 150.1121 (Subcetate 95)
1128 USSR Be 1042.094 (Wysokoje 93)
1129 DR (Pasewalk 70) + 01.03.76
52 1130 PKP *Ty2-172* *+ 17.04.85*
1131 JDZ 33-063
1132 DB + 10.08.57
1133 USSR Kx (Schewtschenko Wrk) + 48
1134 JDZ 33-064 (Tuzla 96)
1135 PKP ? *Ty2 1198 ?*
1136 USSR + 89
1137 USSR Be 1042.096 (Wysokoje 93)
1138 PKP *Ty2-173* *+ 31.08.49*
1139 USSR, Uk
52 1140 PKP *Ty2-174* *+ 02.03.82*
1141
1142 DR (Kamenz 70) + 02.06.76
1143 DR 52 8130 (Elberswalde 70) z 04.88 M
1144 DR + 12.12.69
1145 DR 52 8100 (Berlin Schöneweide 70), Dsp 05.88, M
1146 DR (Dresden 70) + 02.03.78
1147 PKP *Ty2-175* *+ 04.08.70*
1148 USSR (Be 58, Ungeny 78)
1149 USSR (i)
52 1150 USSR (i) (Lesnaja 60)
1151 USSR 1042.613 (Bagrationowsk 93)
1152 DR 52 8040 (Frankfurt/Oder 70) + 16.06.78
1153 SWDE, DB (Offenburg 50) + 15.11.57
1154 DR (Lübbenau 70) + 02.12.75
1155 USSR (Ba 53, La 56, Lw 59) i 65
1156 PKP *Ty2-176* *+ 04.02.91*
1157 PKP *Ty2-177* *+ 08.09.73*
1158 USSR 1042.099 (Zheleznodorozhnyj 93)
1159 DR 52 8113 (Elsterwerda 70) M
52 1160 USSR (Be 58) (63) PKP
1161
1162 PKP *Ty2-178* *+ 04.10.91*
1163 SWDE, DB (Radolfzell 50) + 18.10.54
1164 PKP *Ty2-179* *+ 12.02.86*
1165 DR (Kamenz 70) + 24.10.78
1166 DB ?
1167 PKP
1168 DR (Halle-G 70) + 19.05.76
1169 DR 52 8145 (Berlin Schöneweide 70) Dsp 05.88
52 1170 USSR 1042.100 (Krasnowka 93)
1171 PKP *Ty2-181* *+ 15.05.90*
1172 PKP *Ty2-182* *+ 17.03.90*
1173 PKP *Ty2-183* *+ 02.07.82*
1174 PKP *Ty2-184*
1175 DR 52 8136 (Berlin Schöneweide 70) + 15.08.88
1176 JDZ 33-065
1177 ÖStB, ÖBB + 27.12.62
1178 USSR, (63) MAV 520,058 + 04.64
1179 USSR (Li 47 Ba 53 La 56) + 01.10.74
52 1180 DR (Engelsdorf 70) # + 09.01.76
1181 DR (Elsterwerda 70) + 03.08.73
1182 DB + 07.02.52
1183 USSR (Be 50, MK 53, Kln 58, i 59)
1184 DR (Kamenz 70) + 06.02.78
1185 PKP *Ty2-185* *+ 17.03.90*
1186 USSR (SW 50 i 55) + 01.05.65
1187 DR (Salzwedel 70) + 02.12.75
1188 USSR, (66) JZ 33-336
1189 ÖStB, ÖBB (Selzthal 63) + 04.11.74
52 1190 USSR, (64) JZ 33-311
1191 USSR (Wnz 44)
1192 PKP *Ty2-186* *+ 03.11.89*
1193 PKP *Ty2-187* *+ 22.11.83*
1194 USSR Be 1042.101 (Wysokoje 93)
1195 DR + 20.02.69
1196 PKP *Ty2-188* *+ 08.12.82*
1197 CSD
1198 ÖStB, ÖBB (Selzthal 63) + 27.10.78 M
1199 DR (Pasewalk 70) + 10.06.69
52 1200 PKP *Ty2-189* *+ 28.03.83*
1201 DR (Salzwedel 70) + 07.08.79
1202 PKP *Ty2-190* *+ 22.02.90*
1203 PKP *Ty2-191* *+ 22.08.72*
1204 USSR Be 1042.102 (Wysokoje 93)
1205 DR + 12.12.69
1206 DR (Altenburg 70) + 28.09.79
1207 ÖStB, ÖBB + 17.02.50
1208 PKP *Ty2-192* *+ 09.02.76*
1209 ÖStB, ÖBB + 28.04.77
52 1210 USSR, Uk 1042.103
1211 USSR (Od 50, Len 51, Okt 53, i 57 Kohlenindustrie)
1212 DR (Dresden 70) + 26.02.74
1213 DR (Templin 70) + 06.06.77
1214 PKP *Ty2-193* *+ 17.08.92*
1215 USSR (Ki 47, Wnz 51, SW 53, i 56 Kohlenindustrie)
1216 PKP ? *Ty2-1170 ?*
1217 USSR (W 47, Od 48, OK 53, Od 56, Lw 61)
(62) DR 52 8198 (Templin 70) z 04.88 M
1218 JDZ 33-216 (Tuzla 96)
1219 USSR (BL 50, Be 53, i 57 Kohlenindustrie)
52 1220 PKP ? *Ty2-1171 ?*
1221 DR (Zittau 70) + 07.05.76
1222 PKP *Ty2-404* *+ 20.12.90*
1223 USSR 1042.104 (Kaliningrad 85)
1224 PKP *Ty2-194*

52 1225	DR, (59) BDZ 15.129 (Bodrowo 92)		
1226	USSR, (63) PKP Ty2-1369		+ 15.05.90
1227	ÖStB, ÖBB (Straßhof 63) + 06.12.78		M
1228	PKP	*Ty2-195*	*+ 22.05.91*
1229	USSR (La 56, Mo 60), BDZ 15.138		(Bodrowo 92)
52 1230	PKP	*Ty2-196*	*+ 24.08.88*
1231	PKP		
1232	PKP	*Ty2-198*	*+ 09.90*
1233	USSR	(W 52, Kln 53, Be 57)	i 63
1234	SWDE,DB	(Kirn 50)	+ 15.11.57
1235	USSR	1042.105 (Mamonowo 93)	
1236	DB		+ 10.08.57
1237	PKP	*Ty2-199*	*+ 18.11.86*
1238	ÖStB,	ÖBB (Mürzzuschlag 63)	+ 27.02.75
1239	ÖStB,	ÖBB + 20.06.56 § 985.205	
52 1240	USSR,	(63) CSD 555.0256	+ 69
1241	PKP	*Ty2-200*	*+ 10.12.87*
1242	USSR,	(64) BDZ 15.188 (Asenowo 90)	
1243	PKP	*Ty2-201*	*+ 07.08.73*
1244	USSR	(Kritschew 47), (62) PKP Ty2-1211	
1245	PKP ?	*Ty2-202* ?	
1246	USSR,	(63) CSD 555.0267	+ 70
1247	DB		+ 07.02.52
1248	DR	52 8169 (Stendal 70)	z 10.86 M
1249	USSR,	(63) MAV 520,038	+ 05.71
52 1250	USSR	(Klaipeda 46)	+ 02.51
1251	USSR	(Ki 51, Lw 53)	+ 01.02.72
1252	DR	52 8032 (Frankfurt/Oder 70)	+ 03.07.70
1253	PKP	*Ty2-892*	
1254	DR	52 8181 (Brandenburg 70)	+ 15.08.88
1255	DR	(Zwickau 70)	+ 03.08.73
1256	DB		+ 20.07.59
1257	PKP	*Ty2-893*	*+ 30.07.91*
1258	PKP	*Ty2-894*	*+ 03.05.74*
1259	PKP	*Ty2-895*	*+ 16.12.85*
52 1260	USSR	(Be 46, MK 53)	i 58
1261	PKP	Ty2-896 (Milkowice 91)	+ 18.10.85
1262	PKP	*Ty2-897*	*+ 29.11.90*
1263	USSR,	+ 86 → Vietnam (Hanoi 89)	x 94 ?
1264	USSR,	(63) CSD 555.0272 555.3272	+ 69
1265	PKP	*Ty2-898*	*+ 20.05.87*
1266	DB		+ 25.04.58
1267	USSR	(Kln 51, i 56, Do 61)	i 61
1268	DR	(Roßlau 70)	v 17.12.76
1269	DB		+ 20.07.59
52 1270	DB		+ 20.07.59
1271	DB		+ 20.07.59
1272	USSR	(Kow 45, Lw 53, La 58)	i 58
1273	PKP	*Ty2-899*	*+ 15.05.90*
1274	DR		+ 01.10.68
1275	DR	(Berlin Schöneweide 70)	+ 30.01.80
1276	DB		+ 20.07.59
1277	USSR (i)		
1278	JDZ	33-184	
1279	PKP	*Ty2-900*	*+ 14.01.81*
52 1280	DB		+ 20.07.59
1281	USSR	(OK 53, Od 56, Es 57, i 62 Narva)	
1282	DR	52 8119 (Halle-G 70) Dsp 10.08.88	x 02.92
1283	USSR	1042.107 (Kaliningrad 85)	
1284	DR		+ 26.09.69
1285	DB		+ 20.07.59
1286	USSR,	Uk 1042.108 (Rawa Russkaja 92)	
1287	PKP	*Ty2-901*	
1288	DR	(Bautzen 70)	+ 17.08.81
1289	SWDE, DB	(Freiburg 50)	+ 25.04.58
52 1290	ÖStB,	ÖBB	+ 28.10.54
1291	PKP	*Ty2-902*	*+ 19.11.91*
1292	DR	52 8075 (Wustermark 70)	+ 07.95 M
1293	ÖStB,	ÖBB (Straßhof 63)	+ 20.01.73
1294	USSR,	Uk 1042.109 (Rawa Russkaja 92)	
1295	USSR	(Li 58, Lw 60), (64) JZ	
1296	DB		+ 10.08.57
1297	DR	(Wittenberg 70)	+ 17.03.76
1298	PKP	*Ty2-903*	*+ 15.01.88*
1299	DB		+ 10.08.57

521300	DR, (46) → 52 1447		
1301	USSR	(Be 53, Mo 60), BDZ 15.139	(Bodrowo 92)
1302	PKP	*Ty2-904*	*+ 19.04.78*
1303	USSR	(Kow 45, Lw 53, Ba 54)	+ 01.10.84
1304	PKP	*Ty2-905*	*+ 26.07.89*
1305	USSR		i 76
1306	SNCF	150 Y 1306	
1307	PKP	Ty2-906 (Milkowice 91)	
1308	USSR	(SW 53, Ba 56 , La 56)	+ 01.04.75
1309	USSR	(Li 46, SW 48, S 53)	i 56
521310	USSR,	(63) CSD 555.0252 555.3252	+ 71
1311	PKP	*Ty2-907*	*+ 10.11.87*
1312			
1313			
1314	USSR	(Pillau 46, SW 48)	i 56
1315	DR		+ 12.51
1316	USSR	(01.06.45)	
1317	PKP	*Ty2-908*	*+ 15.01.88*
1318	USSR	(Gomel 47), (64) BDZ 15.264	
1319	USSR	(Li 45, SW 48, Es 56)	+ 01.10.70
521320	USSR	(Li 45, La 48, SW 48, S 53, i 56 Kohlenindustrie)	
1321	USSR	(OK 53, Mo 56), BDZ 15.142	(Bodrowo 92)
1322	USSR,	Uk 1042.110 (Rawa Russkaja 92)	
1323	PKP	*Ty2-909*	*+ 19.12.86*
1324	USSR	(W 51, MK 53, Li 58)	+ 01.01.70
1325	DR	52 8129 (Brandenburg 70)	M
1326	USSR	(SW 49 , La 57, i 58 "Oktoberkohle")	+ 62
1327	ÖStB,	ÖBB + 27.11.57 → Ungarn	
1328	USSR	(Kow 45, Wnz 52, SW 53)	i 56
1329	DR	(Elsterwerda 70)	+ 14.02.74
521330	USSR	1042.111 (Bagrationowsk 93)	
1331	DR	(Frankfurt/Oder 70)	+ 19.11.84
1332	DR	(Eilenburg 70)	+ 15.11.77
1333	USSR,	(63) PKP Ty2-1289	+ 16.08.84
1334	USSR	(Od 56, Li 58), (63) PKP	
1335	USSR		
1336	USSR,	(63) CSD 555.0278 555.3278	+ 72
1337	ÖStB,	ÖBB (Mürzzuschlag 63)	+ 29.12.75
1338	USSR,	(64) BDZ 15.228 (Sindel 90)	
1339	ÖStB,	ÖBB (Selzthal 63)	+ 15.04.75
521340	ÖStB,	ÖBB (Wien Ost 63)	+ 17.02.75
1341	USSR,	(64) BDZ 15.181 (Bodrowo 92)	
1342	USSR	1042.113 (Krasnowka 93)	
1343	USSR,	Be1042.114 (Berestowiza 93)	
1344	PKP	*Ty2-910*	*+ 05.12.73*
1345	USSR,	(64) Jugoslawien 33-503 (Tuzla 96)	
1346	PKP	*Ty2-911*	
1347	USSR,	(63) PKP Ty2-1237	+ 16.07.76
1348	PKP	*Ty2-912*	*+ 31.12.86*
1349	USSR(i)	(Lesnaja 65)	

---

52 1350 - 52 1599

MASCHINENFABRIK ESSLINGEN

1/43 - 5/44

521350	DB		+ 18.10.54
1351	USSR	(La 47, Ba 53, La 56)	+ 01.09.69
1352	PKP	*Ty2-72*	*+ 18.08.89*
1353	USSR	(Osipowitschi 47, Li 57)	+ 01.05.82
1354	DR	(Jüterbog 70)	+ 26.08.76
1355	DR	(Wittenberg 70)	+ 07.08.79
1356	USSR	(Od 45)	
1357	USSR	(Len 51, Okt 53, Be 58)	
	(62) DR	(Altenburg 70)	+ 20.12.79
1358	PKP	*Ty2-73*	*+ 02.08.76*
1359	CSD	555.060, 555.3060	+ 73

52 1360 USSR (SW 48, i 61 Wagen Rep Wrk Darniza) + 66
1361 ÖStB, ÖBB + 30.04.48
1362 JDZ 33-066
1363 USSR, Uk
1364 ÖStB, ÖBB 152.1364, (68) GKB + 69
1365 ÖStB, ÖBB 152.1365 (Graz 63), (69) GKB + 77
1366 ÖStB ÖBB 152.1366 (Wiener Neustadt 63) + 06.12.68
1367 ÖStB ÖBB 152.1367 (Amstetten 63), (73) GKB
1368 USSR (Kln 51) i 59
1369 CCCPö
52 1370 USSR (Ungeny 78, Ungeny 94)
1371 DB + 18.10.54
1372 USSR (St 44)
1373 PKP *Ty2-203* *+ 25.08.88*
1374
1375 DB + 18.10.54
1376 FS, (49) DB + 18.10.54
1377
1378 DB + 19.09.46
1379
52 1380 USSR (Li 45, La 48 , Ba 53, La 56) + 01.05.76
1381 DB + 18.10.54
1382 PKP *Ty2-913* *+ 03.05.86*
1383 USSR (Wnz 44, Lw 47, Od 49, Ba 54, Es 56) + 77
1384 KrV (Szombathely 29.05.45)
1385 CCCPm → CFR 150.1101 ?
1386 USSR (Sloti 46)
1387 KrV (Villach West 45) + 45
1388 ÖStB, ÖBB 152.1388
(Wiener Neustadt 63), (68) → GKB + 73
1389 ÖStB, ÖBB 152.1389 (Amstetten 63) + 20.01.73
52 1390 USSR (Ungeny 78), +86 → Vietnam x 94 ?
1391 USSR (Od 50, Es 57, Okt 61, i 67 Waldindustrie)
1392 USSR (Ungeny 78, Ungeny 94)
1393 DR, (56) BDZ 15.95 (Kalojanowetz 92)
1394 SWDE, DB (Radolfzell 50) + 18.10.54
1395 ÖStB, ÖBB + 01.09.47
1396 ÖStB, ÖBB 152.1396 (Wiener N'stadt 63) +
20.01.73
1397 ÖStB, ÖBB 152.1397 (Wiener N'stadt 63) + 20.10.72
1398 JDZ 33-067
1399 PKP *Ty2-204* *+ 16.06.88*
52 1400 PKP *Ty2-205* *+ 09.08.90*
1401 JDZ 33-068
1402 DR 52 8071 (Halle-G 70) + 09.08.88
1403 SWDE, DB (Radolfzell 50) + 18.10.54
1404 DR (Bautzen 70) + 06.03.79
1405 USSR (SW 47 S 53, i 57 Chemische Industrie)
1406 USSR (Li 47, Ba 53, Li 56) i 64
1407 USSR (Od 48, OK 53, Od 56, i 74 Zuckerfabrik Sasselje)
1408 DR(o) + 08.47
1409 USSR Be 1042.119 (Wysokoje 93)
52 1410 USSR,(63) PKP Ty2-1375 + 15.06.92
1411 PKP *Ty2-206* *+ 08.07.86*
1412 DR (Görlitz 70) + 06.09.84
1413 DR (Halle-G 70) + 30.07.75
1414 USSR (La 47, Ba 53, La 56, Be 58)
(62) DR (Brandenburg 70) + 07.08.79
1415 DR (Görlitz 70) + 12.07.76
1416 USSR 1042.120 (Kaliningrad 85)
1417 USSR 1042.121 (Daugavpils 85)
1418 DR (Salzwedel 70) + 01.03.76
1419 DR (Frankfurt/Oder 70) + 01.07.77
52 1420 DR (Lübbenau 70) + 23.04.75
1421 DR 52 8025 (Frankfurt/Oder 70) + 04.01.90
1422 DR + 20.12.51
1423 DR (Pasewalk 70) z 10.81 M
1424 SWDE, DB + 18.10.54
1425 DR 52 9425 (Senftenberg 70) + 14.11.75
1426 DR 52 9426 (Senftenberg 70) + 29.07.76
1427 USSR (i)
1428 USSR, (62) CSD 555.0246 555.3246 + 68
1429 DR (Hoyerswerda 70) + 24.03.77
52 1430 DB + 18.10.54
1431 DR (Neustrelitz 70) + 19.05.76
1432 SWDE, DB (Radolfzell 50) + 18.10.54
1433 DR (Dresden 70) + 25.10.76
1434 USSR, (63) CSD 555.0222 555.3222 + 70

52 1435 USSR (Lw 46, Ki 48, OK 53, Od 56, i 57 Kohlenind)
1436 DR (Pasewalk 70) v 23.12.77
1437 DR + 26.09.69
1438 DR (Kamenz 70) + 06.03.79
1439 DR (Güsten 70) + 01.03.76
52 1440 DR (Zwickau 70) + 22.11.78
1441 DR(w) + 30.09.46
1442 ÖStB, ÖBB (Wien Ost 63) + 25.10.77 § 9760 032
1443 DR 52 8099 (Halle-G 70) + 13.01.86
1444 USSR (Be 47, Wnz 53, SW53) i 56
1445 USSR (Be 47, SW 48, S 53, i 56 Kohlenindustrie)
1446 DR + 29.08.68
1447 USSR (Kow46, Wnz 52, SW53) i 56
52 1447" (ex 52 1300) → USSR (Be 47, i 57 Kohlenind.)
1448 ÖStB, ÖBB + 15.11.53 § 985.202
1449 DR (Engelsdorf 70) + 07.08.79
52 1450 DR + 20.12.51
1451 DR (Frankfurt/Oder 70) + 12.12.69
1452 DR, (47) → 52 4562" USSR
52 1452" (ex 52 4562) Salzwedel 70 + 19.05.76
1453 USSR (Od 48, OK 53, Kln 55, i 56 Kohlenindustrie)
1454 DR 52 8077 (Falkenberg 70) M
1455 USSR 1042.122 (Zheleznodorozhnyj 93)
1456 USSR, (63) PKP Ty2-1243 + 11.02.72
1457 USSR, Uk 1042.123 (Rawa Russkaja 92)
1458 DR (Eberswalde 70) + 27.12.71
1459 DR (Bautzen 70) + 06.03.79
52 1460 PKP *Ty2-207* *+ 26.07.89*
1461 USSR, (63) PKP Ty2-1393 + 22.04.77
1462 USSR, (66) JZ 33-333
1463 DR (Bautzen 70) v 05.74
1464 PKP *Ty2-208* *+ 29.11.90*
1465 DR + 12.12.69
1466 DR (Frankfurt/Oder 69) v 03.04.79
1467 USSR, (63) CSD 555.0269 555.3269 + 69
1468 DR, (59) BDZ 15.130 (Bodrowo 92)
1469 USSR 1042.124 (Kaliningrad 85) M
52 1470 USSR 1042.125 (Tschernjachowsk 85)
1471 USSR (OK 53, Od 56, Li 58) i 64
1472 DR (Bautzen 70) + 28.09.79
1473 DB + 18.10.54
1474
1475 PKP *Ty2-209* *+ 07.09.88*
1476 USSR 1042.126 (Daugavpils 85)
1477
1478 USSR (SW 50, S 53, Li 58) + 01.10.77
1479 USSR (Be 58) + 88
52 1480 DR + 26.07.68
1481 USSR, (63) PKP Ty2-1401 + 07.90 M
1482 DB + 18.10.54
1483 USSR, (66) JZ 33-335
1484 PKP Ty2-609 (Nysa 84) + 14.08.87
1485 USSR (Ki 44, Od48, OK 53, Es57, i 59, SW 71, i 72)
1486 DB (51) JDZ 33-238
1487 DR (Pasewalk 70) + 27.12.71
1488 USSR (Daugavpils Wrk 50) i 52
1489 DR + 16.09.68
52 1490 DR (Frankfurt/Oder 70) + 04.01.77
1491 DR (Frankfurt/Oder 70) + 20.04.77
1492 USSR (i) (Bulanasch 57)
1493 DB + 18.10.54
1494 PKP *Ty2-211* *+ 14.06.89*
1495 USSR (Be 55, Li 59), (63) PKP Ty2-1367 + 06.09.76
1496 DB + 18.10.54
1497 USSR, (63) PKP Ty2-1253
1498 DR (Frankfurt/Oder 70) + 22.02.79
1499 DR 52 8048 (Wustermark 70) z 07.88 x 93
52 1500 PKP *Ty2-212* *+ 10.09.75*
1501 DR 52 8178 (Stendal 70) + 15.08.88
1502 PKP *Ty2-213* *+ 11.08.78*
1503 DR + 04.51
1504 PKP *Ty2-214* *+ 29.11.90*

52 1505	DR, (53) → 52 7793 → 52 8133 (Brandenburg 70)	M
1505"	Cegielski 887, DR (Berlin Schöneweide 70)	+ 06.03.79
1506	(45) → 52 3638	
1507	(45) → 52 3639	
1508	PKP Ty2-215 (Nysa 87)	+ 15.05.90
1509	USSR (SW 47, MK 48, S 53)	i 57
52 1510	USSR, Uk	
1511	USSR, (63) PKP Ty2-1403	+ 28.12.89
1512	USSR (OK 53, Ba 54, Es 56)	+ 01.04.78
1513		
1514	DR 52 8171 (Stendal 70)	+ 28.03.95
1515	DR (24.04.52 Kesselexplosion Döbeln)	+ 10.11.55
1516	DR (Engelsdorf 70)	+ 20.07.78
1517	CCCPc	
1518	USSR, (64) BDZ 15.238	
1519	USSR, (63) PKP Ty2-1368 (Gdynia 93)	
52 1520	USSR (La 47, Ba 53, La 56)	+ 01.07.80
1521	DR	v 20.12.74
1522		
1523	→ 52 3762	
1524	DB	+ 13.12.50
1525	→ 52 3763	
1526	USSR, (63) MAV 520,026	+ 09.77
1527	USSR, (64) BDZ 15.245 (Bodrowo 92)	
1528	SAAR, DB	+ 25.04.58
1529	SWDE, DB (Ehrang 50)	+ 18.10.54
52 1530	SWDE, DB (Friedrichshafen 50)	+ 18.10.54
1531	PKP *Ty2-216*	*+ 30.11.86*
1532	DB	+ 13.12.50
1533	DB	+ 18.10.54
1534	DB	+ 18.10.54
1535	DB	+ 18.10.54
1536	DB	+ 18.10.54
1537	USSR (SW 48)	i 56
1538	DB	+ 18.10.54
1539	DR (Elsterwerda 70)	+ 17.02.76
52 1540	DB	+ 18.10.54
1541	DB	+ 18.10.54
1542	DB, (51) JDZ 33-264	
1543	DB	+ 18.10.54
1544	DB	+ 18.10.54
1545	FS, (48) DB	+ 18.10.54
1546	DB	+ 18.10.54
1547	CSD, (58) BDZ 15.121	
1548	CSD 555.061 ?	
1549	CSD 555.0601, 555.0185, 555.3185 #	+ 70
52 1550	DB	+ 18.10.54
1551	DB	+ 18.10.54
1552	DB	+ 18.10.54
1553	USSR (Dno 81)	+ 88
1554	USSR 1042.130 (Kaliningrad 85)	
1555	USSR (Kow 50, Lw 53, La 57)	
	(62) DR (Eberswalde 70)	+ 06.02.78
1556	ÖStB, ÖBB (Wien Ost 63)	+ 01.02.73
1557	DB	+ 18.10.54
1558	USSR (Li 47, Ba 53, Es 56)	i 58
1559	DB	+ 18.10.54
52 1560	ÖStB, ÖBB (Wels 63)	+ 25.05.71
1561	FS, (47) DB	+ 18.10.54
1562	DR (Frankfurt/Oder 70)	+ 19.05.76
1563	DB	+ 18.10.54
1564	USSR (Len 51, Okt 53, i 57 Kalinin)	
1565	DB	+ 18.10.54
1566	PKP *Ty2-217*	*+ 30.04.73*
1567	DB	+ 18.10.54
1568	USSR, (62) CSD 555.0213 555.3213	+ 69
1569	DR 52 8011 (Cottbus 70)	+ 19.08.88
52 1570	USSR (Li 47, Ba 53, Li 56)	+ 01.11.77
1571	(CSD 555.0135) ?	
1572	DR 52 8018 (Cottbus 70)	x 03.90
1573	USSR (Okt 53, i 57 Tambow)	+ 67
1574	DR (Bautzen 70)	+ 14.07.75
1575	SWDE, DB (Freudenstadt 50)	+ 18.10.54
1576	DB	+ 18.10.54
1577	USSR (SW 50)	i 56
1578	USSR (Od 56, Li 58, i 76 Riga)	
1579	PKP *Ty2-914 ?*	

52 1580	DR 52 9580 (Senftenberg 70)	+ 25.10.76
1581	USSR 1042.131 (Bagrationowsk 93)	
1582	USSR (Gomel 47), (64) JZ 33-309	
1583	DR (Bautzen 70)	+ 07.08.79
1584	DB	+ 18.10.54
1585	*(PKP Ty2-915) ?*	
1586	DB, (51) JDZ 33-265	
1587	DB	+ 18.10.54
1588	DB	+ 18.10.54
1589	DB, (51) JDZ 33-261	
52 1590	DB	+ 18.10.54
1591	ÖStB, ÖBB (Wiener Neustadt 63)	+ 28.04.77
1592	DB	+ 18.10.54
1593	SNCF 150 Y 1593	
1594	ÖStB, ÖBB + 10.08.59 § 985.210	
1595	DB (München Hbf 50)	+ 18.10.54
1596	CSD 555.086 (59) BDZ 15.123	
1597	DB	+ 18.10.54
1598	DB	+ 18.10.54
1599	DB	+ 18.10.54

---

52 1600 - 52 1738

GRAFFENSTADEN

6/43 - 10/44

---

52 1600	PKP Ty2-218 (Jaworzyna 84)	+ 19.11.91
1601	PKP	
1602	DR (Bautzen 70)	v 06.10.76
1603	DR 52 8172 (Brandenburg 70)	+ 19.08.88
1604	DR 52 8092 (Leipzig West 70)	Dsp 01.90 M
1605	DR 52 8009 (Cottbus 70)	z 12.85 M
1606	DB, (52) Frankreich	
1607	USSR, (63) CSD 555.0281 555.3281	+ 70
1608	PKP ? *Ty2-220 ?*	
1609	DR (Stralsund 70)	+ 17.02.76
52 1610	ÖStB, ÖBB (Straßhof 63)	+ 16.12.74
1611	USSR (Ba 53, Li 56, i 75 Gusew)	
1612	USSR (Be 53)	+ 81
1613	PKP *Ty2-221*	*+ 20.02.76*
1614	USSR 1042.132 (Jelgava 85)	+ 87
1615	DR (Seddin 70)	+ 27.12.71
1616	USSR, (63) PKP Ty2-1250	
1617	DR 52 8036 (Hoyerswerda 70)	z 01.88 M
1618	USSR 1042.133 (Mamonowo 93)	
1619	USSR (Karolina Bugas 46)	
52 1620	USSR (Kx, Podgorodnaja 46)	
1621	FS, (48) DB, (52) Frankreich	
1622	USSR, (63) MAV 520,027	+ 01.78
1623	DR (Frankfurt/Oder 70)	+ 17.01.72
1624	DB, (52) Frankreich	
1625	USSR, (63) MAV 520,059	+ 05.77
1626	DR (Eilenburg 70)	+ 19.05.76
1627	FS, (49) DB, (52) Frankreich	
1628	SWDE, (52) Frankreich	
1629	DB, (52) Frankreich	
52 1630	DB	
1631	USSR,Be 1042.134 (Brest Wostok 93)	
1632	SWDE, (52) Frankreich	
1633	DR 52 8123 (Eberswalde 70)	M
1634	DB, (52) Frankreich	
1635	DR (Halle-G 70)	+ 17.01.72
1636	DR (Bautzen 70)	v 17.12.75
1637	SNCF 150 Y 1637	
1638	DB, (52) Frankreich	
1639	DB, (52) Frankreich	
52 1640	DB, (52) Frankreich	
1641	DB, (52) Frankreich	
1642	USSR (OK 53, Ba 55, Li 56)	+ 71
1643	PKP *Ty2-222*	*+ 23.02.90*
1644	USSR, (64) JZ 33-312	

52 2005 DB (Mainz-Bischofsh. 50) ,+ 01.06.53
2006 DR(w), (45) USA
2007 PKP
2008 DR (Kamenz 70) + 01.02.77
2009 DR, (59) BDZ 15.134 ?
52 2010 DR + 59 Rahmen → 52 192
2011 DR (Kamenz 70) v 17.12.76
2012 DB (Kirchweyhe 50) + 14.11.52
2013 DB (Mainz-Bischofsh. 50) + 01.06.53
2014 DB + 53, → Heizlok, (Weiden 56)
2015 DB (Mainz-Bischofsh. 50) + 01.06.53
2016 DB + 01.06.53
2017 DB (Mainz-Bischofsh. 50) + 01.06.53
2018 DB (Minden 50) + 01.06.53
2019 DB (Minden 50) + 01.06.53
52 2020 DB (Mainz-Bischofsh. 50) + 01.06.53
2021 DB (Mainz-Bischofsh. 50) + 01.06.53
52 2021" ex 52 2821 USSR (SW50, Od 52, Ok 53, Od 56, Be 59), (63) PKP Ty2-1258 (Elk 90) + 15.03.92
2022 DB (Mainz-Bischofsh. 50) + 01.06.53
2023 DB (Minden 50) + 09.11.53
2024 DB + 01.06.53
2025 DB + 53, → Heizlok, (Nürnberg 53), Zerlegt 61
2026 DB (Kirchweyhe 50) + 09.11.53
2027 DB (Kirchweyhe 50) + 01.06.53

52 2090 - 52 2893

HENSCHEL

52 2090 - 2624 1/43 - 12/43
52 2625 - 2868 1/44 - 12/44
52 2869 45
52 2870 11/44
52 2871 9/45
52 2872 11/44
52 2873 1/45
52 2874 - 2876 11/44 - 12/44
52 2877 10/45
52 2878 - 2882 11/44
52 2883 - 2887 1/45 - 3/45
52 2888 - 2893 5/45 - 7/46

---

52 2090 PKP *Ty2-74* *+ 04.11.89*
2091 PKP *Ty2-75* *+ 23.04.76*
2092 USSR 1042.147 (Bagrationowsk 93)
2093 DR (Dresden 70) z 23.06.81 M
2094 USSR (Li 50, Ba 53, Es 56) i 69
2095 PKP
2096 USSR 1042.148 (Krasnowka 93)
2097 BDZ 15.09
2098 SWDE, DB (Freiburg 50) + 18.10.54
2099 USSR Be 1042.149 (Luninez 93)
52 2100 PKP *Ty2-77* *+ 87*
2101 DB + 18.10.54
2102 USSR (Od 44-48) Kessel →TE-8022 ?
2103 USSR (Lw 46, Ki 48, Lw 49, Kln 51, Okt 61) + 01.06.77
2104 PKP *Ty2-78* *+ 20.09.88*
2105 JDZ 33-069
2106 PKP *Ty2-79* *+ 11.05.78*
2107 DB + 18.10.54
2108 PKP *Ty2-80*
2109 SWDE, DB (Freudenstadt 50) + 18.10.54
52 2110 DR(w) + 20.09.48
2111 USSR (BL 47, Li 48, Okt 53) i 73
2112 PKP *Ty2-81*
2113 PKP *Ty2-82* *+ 20.06.88*
2114
2115
2116 PKP *Ty2-83* *+ 05.06.87*
2117 CCCPd
2118 USSR (Mo 56, La 58, Be 63), → Pi (Szczakowa 91) M
2119 USSR (Kow 46) i 52
52 2120 PKP *Ty2-84* *+ 23.02.90*
2121 ÖStB, ÖBB 152.2121 (Knittelfeld 63) + 22.11.72
2122 CCCPm, USSR (Lw 47)
2123 PKP *Ty2-85* *+ 14.07.86*
2124 JDZ 33-070
2125 CCCPö
2126 USSR + 88
2127 PKP *Ty2-223* *+ 30.05.84*
2128 USSR 1042.151 (Radviliskis 85) + 86
2129 DR (Berlin Schöneweide 69) + 20.04.77
52 2130 USSR 1042.152 (Tapa 85) + 88
2131 PKP *Ty2-86* *+ 19.11.91*
2132 USSR (Okt 53, i 75 Kotlas)
2133 DR(w) + 20.09.48
2134 PKP *Ty2-87* *+ 23.03.70*
2135 PKP *Ty2-224* *+ 13.04.79*
2136 USSR (Od 50, Okt 53, i 57 Waldindustrie)
2137 DR(o) + 46
2138 PKP *Ty2-225*
2139 DB + 18.10.54
52 2140 USSR (Ungeny 78)
2141 USSR (Kow 45 ,IF 46, Od 53, ...), (63) PKP
2142 USSR 1042.154 (Vilnius 85) + 89
2143 USSR (Od 44, BL 47, Be 53, La 58) + 01.12.75
2144 USSR 1042.155 (Gusew 92)

52 2145 USSR (Lw 53), (64) BDZ 15.222
2146 USSR (SW 50) i 56
2147 USSR (Be 47, SW 48, S 53) + 09.59
2148 USSR (Od 44) i 45
2149 PKP *Ty2-226* + *29.12.86*
52 2150 USSR (W 45, La 47)
2151 DR (Berlin Schöneweide 70) + 11.04.74
2152 DR, (56) BDZ 15.100 (Cervan Brjag 90)
2153
2154 DB + 24.10.52
2155 PKP *Ty2-228* + *90*
2156 SWDE, DB (Freudenstadt 50) + 18.10.54
2157 PKP *Ty2-229* + *74*
2158 DB + 18.10.54
2159 USSR Be 1042.156 (Berestowiza 93)
52 2160 DB + 18.10.54
2161 DR (Frankfurt/Oder 70) + 01.07.77
2162 USSR (SW 50, Es 56) + 01.12.70
2163 DR (Berlin Schöneweide 70) + 12.02.80
2164 PKP *Ty2-230* + *10.04.91*
2165 DR (Angermünde 70) + 03.08.73
2166 USSR (St 44) PKP *Ty2-231*
2167 DR (Cottbus 70) + 20.12.79
2168 USSR (Ki 44)
2169 DB + 18.10.54
52 2170 USSR, (63) MAV 520,060 + 02.77
2171 USSR, (63) CSD 555.0268 + 68
2172 USSR (Kln 51, Okt 55, Be 58), (63) PKP
2173 PKP *Ty2-232* + *01.89*
2174 USSR (Kow 46, Lw 49), (64) BDZ 15.176
2175 PKP *Ty2-233* + *18.08.89*
2176 USSR (Be 50, Kln 54, Ms 61, Lw 64) i 75
2177 DR(w), 9/45 DR(o), 52 8127 x 10.91
2178 PKP *Ty2-234* + *89*
2179 USSR (Wjerbka 89)
52 2180 DR 52 8118 (Frankfurt/Oder 70) z 03.09.87 M
2181 USSR (SW 44, Od 49, Es 57, i 59) + 72
2182 USSR, (64) BDZ 15.197 (Kalojanowetz 92)
2183 USSR (Ki 44)
2184 USSR (Li 45, Ba 53, La 56, Mo 60), BDZ 15.141
2185 PKP *Ty2-235* + *89*
2186 USSR (SW 44, Li 45, Be 45, MK 53, i 59) + 65
2187 PKP *Ty2-236* + *04.09.78*
2188 USSR (Be 58, Ungeny 94)
2189 USSR, (63) MAV 520,061 + 03.67
52 2190 PKP *Ty2-237*
2191 USSR (SW 44, Be45, St 52, Wnz 52, SW 53) i 56
2192 DR (Berlin Schöneweide 69) + 12.12.69
2193 USSR (Wnz 44, Od 48, OK 53) i 55
2194 USSR 1042.159 (Mamonowo 93)
2195 DR (Frankfurt/Oder 70) M
2196 USSR 1042.160 (Vilnius 85) + 88
2197 USSR (SW44, Be 45, La 50, Be 58),
(63) PKP Ty2-1345 + 17.03.77
2198 USSR, (63) MAV 520,062 + 06.78
2199 DR 52 8069 (Lübbenau 70) z 03.88 M
52 2200 DB + 18.10.54
2201 USSR, (62) CSD 555.0236, 555.3236 + 68
2202 USSR (Li 45, La 47, Be 58)
(62) DR (Cottbus 70) v 07.07.76 M
2203 DB + 18.10.54
2204 USSR, (63) CSD 555.0264, 555.3264 + 68
2205 PKP
2206 PKP *Ty2-239* + *18.04.91*
2207 USSR 1042.161 (Tschernjachowsk 93)
2208
2209 DB, (51) JDZ 33-256
52 2210 JDZ 33-071
2211
2212 DB + 18.10.54
2213 DB, (51) JDZ 33-258
2214 PKP *Ty2-240* + *10.08.90*
2215 PKP Ty2-1186 (Milkowice 91)
2216 USSR (Kow 45, Be 47, SW 48) i 56
2217 USSR, (63) MAV 520,048 + 09.71
2218 USSR 1042.162 (Radviliskis 85) + 88
2219 CSD 555.098 + 69

52 2220 ÖStB, ÖBB + 15.05.50
2221 USSR (Ki 44, OK 53, Mo 56, La 58, i 69 Brauerei "Saku")
2222 DB + 18.10.54
2223 USSR, (63) MAV 520,063 + 06.78
2224 USSR (Li 45, Ki 48, Lw 53, Mo 56, Be 60), (62) PKP
2225 PKP *Ty2-241* + *18.02.87*
2226 DR (Frankfurt/Oder 70) + 30.09.71
2227 DB + 18.10.54
2228 DB + 18.10.54
2229 DB + 18.10.54
52 2230 PKP *Ty2-242* + *17.03.90*
2231 DB + 18.10.54
2232 USSR 1042.163 (Zheleznodorozhnyj 93)
2233 PKP *Ty2-243* + *04.04.85*
2234 USSR (Li 49, Ba 53, Li 56, Be 58), (63) PKP Ty2-1323
2235 JDZ 33-186
2236 DB + 18.10.54
2237 USSR, (64) JZ 33.272 ?
2238 USSR (Li 45, Ba 53 Li 56) + 77
2239 USSR (Lw 53), (63) MAV 520,076 + 05.77
52 2240 PKP *Ty2-244* + *15.09.89*
2241 USSR (Od 44)
2242 DB + 18.10.54
2243
2244 DB + 18.10.54
2245 DB + 18.10.54
2246 USSR (SW 45)
2247 USSR 1042.164 (Krasnowka 93)
2248 USSR (Lw 48, Od 52, OK 53, Od 56, Li 58) + 73
2249 PKP *Ty2-245* + *19.01.72*
52 2250 PKP *Ty2-246* + *04.07.91*
2251 USSR 1042.165 (Radviliskis 85) + 89
2252 USSR (Wnz 44, Lw 47, Od 48, OK 53, Od 56) i 56
2253 CSD
2254 USSR 1042.166 (Jelgava 90)
2255 SWDE, DB (Freiburg 50) + 18.10.54
2256 DB + 02.11.56
2257 SWDE, DB (Friedrichshafen 50) + 18.10.54
2258 JDZ 33-213
2259 PKP *Ty2-248*
52 2260 USSR Be 1042.167 (Brusgi 93)
2261 PKP *Ty2-249* + *23.07.92*
2262 SWDE, DB (Kirn 50) + 18.10.54
2263 PKP Ty2-430 (CN 91) + 23.12.88
2264 PKP *Ty2-251* + *30.04.87*
2265 PKP *Ty2-252*
2266 USSR (Od 46)
2267 USSR (St 44) PKP *Ty2-253* + *28.12.89*
2268 USSR (St 44) PKP *Ty2-254* + *16.04.86*
2269 PKP *Ty2-255* + *19.08.85*
52 2270 DR (Elsterwerda 70) + 21.02.77
2271 USSR 1042.168 (Radviliskis 85) + 88
2272 SWDE, DB (Freudenstadt 50) + 18.10.54
2273
2274 USSR (St 44, NW45, Lw46, Be46 , SW48, La 57) i 58
2275 USSR 1042.169 (Tapa 85) + 88
2276 USSR (Od 44) i 45
2277 USSR, (63) PKP Ty2-1387 + 04.92 M
2278 JDZ 33-187
2279 PKP Ty2-256 (Milkowice 91) + 22.04.88
52 2280 BDZ 15.10 (Kalojanowetz 92)
2281 DR (Elsterwerda 70) + 14.05.79
2282 PKP *Ty2-257*
2283 USSR, (63) PKP
2284 DB + 18.10.54
2285 PKP *Ty2-258* + *18.08.76*
2286 USSR (Od 44, Wnz 48, Lw 50, Od 52, OK 53) i 56
2287 USSR, (63) MAV 520,002 + 06.76
2288 USSR 1042.170 (Kaliningrad 85) + 90
2289 DR 52 8143 (Cottbus 70) z 07.87 x 05.92
52 2290 PKP *Ty2-259* + *26.07.89*
2291 USSR (Li 45, La 48, Ba 53, La 56,
i 84 Zuckerfabrik Dobele)
2292 JDZ 33-072
2293 NSB 2293 + 11.70
2294 PKP *Ty2-260* + *04.02.91*

52 2295 DR 52 8086 (Halle-G 70) z 04.87 M
2296 USSR (Kow 50, Wnz 52, SW 53) i 56
2297 USSR Be 1042.171 (Grodno 93)
2298 DR(o) + 46
2299 USSR (Be 48, W 51, Kln 53, i 56 Kohlenindustrie)
52 2300 PKP *Ty2-920* *+ 23.04.76*
2301 PKP
2302 USSR (Lw 53), (63) MAV 520,039 + 03.71
2303 USSR (Wnz 44)
2304 USSR (Li 47, Ba 53, Li 56) + 01.10.71
2305 USSR (Wnz 45)
2306 PKP *Ty2-261* *+ 27.12.82*
2307 DB + 18.10.54
2308 USSR (Luninez 90)
2309 USSR (Kow 46)
52 2310 USSR, (64) BDZ 15.267 (Kalojanowetz 92)
2311 PKP Ty2-262 (Zbaszynek 92)
2312 DR 52 8096 (Berlin Schöneweide 70) v 07.92 #
2313 DB + 18.10.54
2314 PKP *Ty2-263* *+ 23.05.91*
2315 USSR (Lw 52), (63) MAV 520,015 + 04.76
2316 USSR (La45, ..., SW48, Od 52, OK 53, Od 56) i 56
2317 USSR, (62) CSD 555.0225, 555.3225 + 67
2318 CCCPö
2319 USSR (Gomel 47) i 57
52 2320
2321 USSR (Od 44) i 45
2322
2323 DR(o) + 46
2324 DR (Neustrelitz 70) + 15.03.73
2325 ÖStB, ÖBB (Knittelfeld 63) + 21.03.75
2326 USSR (Li 46, La 48, Ba 53, Es 56) + 01.08.75
2327
2328 PKP *Ty2-264* *+ 18.03.82*
2329
52 2330 DR (Elsterwerda 70) + 08.02.74
2331 USSR (Lw 53), (63) MAV 520,064 + 04.78
2332 PKP *Ty2-265* *+ 28.12.89*
2333 DR + 29.10.69
2334 DR(w) + 20.09.48
2335 ÖStB, ÖBB + 25.01.61
2336 PKP Ty2-266 (Bydgoszcz 93)
2337 SWDE, DB (Worms 50) + 18.10.54
2338
2339 DB + 18.10.54
52 2340 USSR 1042.173 (Krasnowka 93)
2341 BDZ 15.11
2342 DB + 18.10.54
2343 NSB 2343 + 04.70
2344 PKP *Ty2-922*
2345 DB + 18.10.54
2346 PKP *Ty2-267*
2347 USSR (Königsberg Wrk 46) + 48
2348 PKP *Ty2-268* *+ 21.02.88*
2349 ÖStB KrV + 01.09.47
52 2350 DB, (52) JDZ
2351 DR (Engelsdorf 70) + 17.02.76
2352 DR (Bautzen 70) + 17.02.76
2353 PKP *Ty2-269* *+ 23.07.92*
2354 USSR (BL 48 , W 51, Kln 53, i 56 Kohlenindustrie)
2355 DR (Berlin Schöneweide 70) + 26.08.76
2356 PKP *Ty2-270* *+ 24.08.80*
2357 USSR (Mo 61), (63) MAV 520,040 + 05.71
2358 DB + 18.10.54
2359 ÖStB, ÖBB + 14.09.57 → Ungarn
52 2360 DB + 18.10.54
2361 USSR (Od 56, Li 58),
(63) PKP Ty2-1381 + 05.10.90 → Dsp (Szczecinek 96)
2362 DR (Karl-Marx-Stadt 70) v 03.11.75
2363 USSR + 86 → Vietnam x 94 ?
2364 PKP *Ty2-271* *+ 07.06.85*
2365 DR + 20.12.51
2366 USSR (Lw 48) + 69
2367 DB + 18.10.54
2368 USSR (Li 56, Lw 60), BDZ 15.248
2369 PKP *Ty2-272* *+ 26.05.76*

52 2370 PKP *Ty2-273* *+ 13.03.87*
2371 PKP *Ty2-274* *+ 30.04.87*
2372 PKP *Ty2-275* *+ 07.10.88*
2373 USSR (Orscha 46) BDZ 15.242 (Kalojanowetz 92)
2374 ÖStB, ÖBB (Straßhof 63) + 25.07.77 § 985.227
2375 DR(o) + 47
2376 USSR (Okt 53, Be 60), (63) PKP Ty2-1254 + 30.03.87
2377 USSR, (66) JZ 33-339 M
2378 USSR (St 44) + 48
2379 USSR (Kow 46, Lw 53, Ba 54, Li 56) i 58
52 2380 PKP *Ty2-276* *+ 13.06.89*
2381 USSR 1042.175 (Radviliskis 85) + 89
2382 DR(o) + 47
2383 USSR (Lw 44, NK 45, Be 46), (63) PKP Ty2-1332 + 90
2384 DB + 18.10.54
2385 PKP *Ty2-277* *+ 25.09.92*
2386 DR (Wittenberg 70) v 01.01.75
2387 DR 52 8072 (Cottbus 70) z 02.88 M
2388 DB + 18.10.54
2389 DR(o) + 46
52 2390 USSR (La 47, ... , Mo 60), (63) MAV 520,009
(Bw Fonyod 70, Kesselexplosion) + 06.70
2391 PKP *Ty2-278* *+ 20.09.88*
2392 USSR, (64) BDZ 15.227
2393 USSR (Okt 53, Li 58), (63) PKP Ty2-1405 + 08.78
2394 USSR (Ungeny 78, Ungeny 94)
2395 DB + 18.10.54
2396 USSR 1042.177 (Kretinga 89) → LG
2397 ÖStB, ÖBB + 20.06.56 § 985.206
2398 USSR 1042.178 (Bagrationowsk 93)
2399 USSR # (64) BDZ 15.247 (Asenowo 92)
52 2400 PKP *Ty2-279* *+ 23.02.90*
2401
2402 PKP *Ty2-280* *+ 12.03.86*
2403 PKP *Ty2-281* *+ 24.11.88*
2404 DR (Dresden 70) + 20.05.81
2405 PKP *Ty2-282* *+ 04.11.89*
2406 CCCPd
2407 USSR (Od 49, OK 53, Ba 55, La 56) + 01.03.72
2408 USSR Uk (Wjerbka 89)
2409 DR (Wittenberg 70) + 24.10.78
52 2410 DR + 17.03.70
2411 USSR (Okt 53, Li 58), (63) PKP Ty2-1386
2412 DR (Wittenberg 70) + 25.10.76
2413 USSR 1042.180 (Daugavpils 85) + 90
2414 USSR (Li 47, Ba 53, La 56) + 01.04.76
2415 DB + 18.10.54
2416 DR (Halle 70) + 06.09.75
2417 USSR (BL47, Be 53), (63) PKP Ty2-1261 + 10.05.76
2418 USSR (Okt 53, i 57 Kohlenindustrie)
2419 USSR (SW48, Od 57, Lw 62), (63) CSSR
52 2420 USSR (SW 50, Es 56) i 58
2421 GySEV 501, (52) DB + 18.10.54
2422 FS, (49) DB + 18.10.54
2423 PKP *Ty2-283* *+ 25.03.81*
2424 ÖStB, ÖBB + 01.12.47
2425 ÖStB, ÖBB (Mürzzuschlag 63) + 12.11.75
2426 USSR (Kln 51, Okt 61) + 01.08.80
2427 ÖStB, ÖBB + 20.11.54
2428 ÖStB, ÖBB (Knittelfeld 63) + 25.10.77 § 985.229
2429 BDZ 15.12 (Asenowo 92)
52 2430 DR + 27.12.71
2431 DR (Salzwedel 70) v 28.07.75
2432 PKP *Ty2-284* *+ 25.03.88*
2433 USSR (Lw 48 , Li 59) + 01.09.72
2434 USSR 1042.181 (Vilnius 85) + 89
2435 DB + 18.10.54
2436 ÖStB, ÖBB (Knittelfeld 63) + 26.07.77 M
2437 ÖStB, ÖBB (St Valentin 63) + 29.11.77
2438 USSR, (64) BDZ 15.220 (Kalojanowetz 92)
2439 FS, (48) DR(w) + 20.05.51
52 2440 ÖStB, ÖBB (Straßhof 63) + 01.02.73
2441 JDZ 33-073
2442 ÖStB, ÖBB + 20.11.54
2443 SAAR, (57) DB + 11.05.60
2444 PKP *Ty2-285* *+ 12.04.78*

52 2445 PKP *Ty2-286* + *13.04.79*
2446 USSR Uk 1042.182
2447 USSR (Daugavpils Wrk 50) i 50
2448 DB + 18.10.54
2449 PKP *Ty2-287* + *11.08.78*
52 2450 DB + 18.10.54
2451 USSR (Lw 58), (63) MAV 520,086 # + 05.84
2452 USSR Be 1042.183 (Wysokoje 93)
2453 DR 52 8174 (Stendal 70) M
2454 USSR (Li 47, Len 53, Okt 53, i 57 Kohlenindustrie)
2455 DR 52 8087 (Frankfurt/Oder 70) + 07.95 M
2456 PKP *Ty2-288* + *01.06.74*
2457
2458 ÖStB, ÖBB (Wien Ost 63) + 29.11.77
2459 DB + 18.10.54
52 2460 DB + 18.10.54
2461 DR + 30.04.51
2462 DB + 18.10.54
2463 CSD
2464 CSD 555.0103, (60) → Jugoslawien 33-501 (Ahmici 96)
2465 DR (Kamenz 70) + 25.08.78
2466 USSR (Lw 48), (63) MAV 520,065 + 11.72
2467 DR 52 8062 (Hoyerswerda 70) z 04.88 M
2468 JDZ 33-188
2469 DR + 20.12.51
52 2470 PKP *Ty2-289* + *19.10.74*
2471 USSR, (63) CSD 555.0274, 555.3274 + 68
2472 DR (Brandenburg 70) + 06.09.75 v 17.10.75
2473 DR (Berlin Schöneweide 70) + 29.07.76
2474 DR (Stralsund 70) + 03.08.73
2475 PKP *Ty2-290* + *26.10.77*
2476 USSR Uk 1042.184 (Rawa Russkaja 92)
2477 DR (Frankfurt/Oder 70) + 19.05.76
2478 PKP *Ty2-291* + *17.07.87*
2479 PKP *Ty2-292* + *10.07.86*
52 2480 DR(o) + 46
2481 USSR (Be 47, i 57 Kohlenindustrie) + 01.02.73
2482 USSR (OK 53, Od 56, La 58, Be 58), (63) PKP
2483 DR (Frankfurt/O 70) + 03.07.70
2484 DB + 18.10.54
2485 DR(o), 45 → DR(w) + 18.10.54
2486 USSR (SW 47, S 53, SW 53 , Es 56) + 01.12.71
2487 PKP *Ty2-293* + *29.11.90*
2488 PKP *Ty2-294* + *74*
2489 PKP *Ty2-295* + *74*
52 2490 DB + 18.10.54
2491 DR (Karl-Marx-Stadt 70) + 20.02.75
2492 DR (Bautzen 70) + 22.02.79
2493 DR (Jüterbog 70) + 19.08.77
2494 DR (Bautzen 70) + 31.01.79
2495 CSD 555.063, (59) BDZ 15.116 (Asenowo 90)
2496 DB + 18.10.54
2497 USSR Be 1042.185 (Berestowiza 93)
2498 → CFR 150.1117 ?
2499 USSR, (63) MAV 520,016 + 09.67
52 2500
2501 DR 52 8124 (Eberswalde 70) v 05.92 #
2502 PKP *Ty2-296* + *12.11.85*
2503 USSR (Kow 47, Ki 48, OK53) i 55
2504 USSR, (62) CSD 555.0243, 555.3243 + 68
2505 USSR (Luninez 90) M
2506 PKP Ty2-297 (Nysa 87)
2507 PKP *Ty2-298* + *14.06.89*
2508 DB + 18.10.54
2509 PKP *Ty2-299* + *31.03.69*
52 2510 PKP *Ty2-300* + *07.06.69*
2511 DB + 18.10.54
2512 BDZ 15.13 (Asenowo 92)
2513 PKP *Ty2-301* + *02.12.90*
2514 PKP *Ty2-302* + *13.09.88*
2515 USSR (Li 47, Ba 53, Li 56), (63) PKP Ty2-1400 ?
2516 PKP *Ty2-303* + *02.89*
2517 PKP *Ty2-304* + *05.10.90*
2518 USSR (Li 47, Ba 53, Li 56), (63) PKP Ty2-1265 + 90
2519 USSR (Okt 53, Li 58), (63) PKP

52 2520 PKP *Ty2-305*
2521 USSR (Okt 53, Be 62), → Pi (Szczakowa 91)
2522 USSR Be 1042.187 (Wysokoje 93)
2523 DR (Frankfurt/Oder 70) + 19.05.76
2524 PKP *Ty2-306*
2525 CSD 555.0108 + 70
2526 USSR, (63) PKP Ty2-1260 (Nysa 85) + 13.08.87
2527 DB + 18.10.54
2528 DR (Bautzen 70) + 06.02.78
2529 DB + 18.10.54
52 2530 CSD ?
2531 CSD 555.0109
2532 DB + 18.10.54
2533 DB + 18.10.54
2534 USSR (Be 47, Wnz 53, SW53, Ba 56, Li 56) + 74
52 2534" USSR (Lw 51), (64) BDZ 15.218
2535 USSR (Be 50), (63) PKP Ty2-1315 + 25.10.74
2536 USSR (Od 50, OK 53, Od 56, Li 58, i 75 Tartu)
2537 USSR (Li 45, La 48, SW 48) i 55
2538 USSR (Kow 50, Lw 53, Ba 54, Li 56, i 73 Akmene)
2539 DR (Elsterwerda 70) + 24.10.75
52 2540 PKP *Ty2-307* + *15.01.88*
2541 DB + 18.10.54
2542 DR 52 8024 (Frankfurt/Oder 70) + 12.11.87
2543 DR 52 9543 (Senftenberg 70) + 10.04.79
2544 DR (Frankfurt/Oder 70) + 26.09.69
2545 DR (Wustermark 70) + 03.07.79
2546 DB + 18.10.54
2547 USSR(i) (Lesnaja 70)
2548 DR (Eilenburg 70) + 19.05.76
2549
52 2550 PKP *Ty2-308* + *05.06.87*
2551 DR 52 8053 (Frankfurt/Oder 70) z 03.90 M
2552 DB + 18.10.54
2553 PKP *Ty2-309* + *07.02.78*
2554 DR (Wittenberg 70) + 02.06.76
2555 ÖStB, ÖBB (Graz 63) + 10.04.69
2556 PKP *Ty2-310* + *25.03.75*
2557 DR (Zittau 70) + 06.09.75
2558 DB + 18.10.54
2559 PKP *Ty2-311* + *29.11.90*
52 2560 DR 52 8007 (Eberswalde 70) z 07.87 M
2561 DR (Cottbus 70) + 05.02.81
2562 DB + 18.10.54
2563 USSR + 85
2564 USSR 1042.189 (Daugavpils 85) + 89
2565 DR (Eilenburg 70) + 03.07.79
2566 JDZ 33-074
2567 PKP *Ty2-923* + *16.05.75*
2568
2569 DB + 18.10.54
52 2570 NSB 2570 (Trondheim 70) + 11.70
2571 DR (Bautzen 70) + 14.11.75
2572 NSB 2572 + 04.70
2573 USSR (Be 50, Lw 59), (63) MAV 520,077 + 08.73
2574 USSR (Kln 51) i 56
2575 DR (Eberswalde 70) + 03.08.73
2576 DR 52 9576 (Senftenberg 70) + 09.01.76
2577 PKP *Ty2-312* + *11.04.86*
2578 CCCPm
2579 PKP *Ty2-313* + *05.04.91*
52 2580 PKP *Ty2-314* + *17.03.90*
2581 CCCPd
2582 DR (Seddin 70) + 25.08.78
2583 PKP *Ty2-315* + *12.02.86*
2584 USSR (Li 47, Len 51, Okt 53) i 57
2585 USSR (Be 47) i 53
2586 USSR 1042.190 (Tschernjachowsk 85) + 89
2587 SAAR, (57) DB + 20.07.59 (Heizlok Bw Trier 71)
2588 USSR Uk (Wjerbka 89)
2589 USSR (Kln 51, Li 58, Be 58), (63) PKP
52 2590 USSR 1042.192 (Riga 85)
2591 DR (Brandenburg 70) + 02.12.75
2592 DB + 18.10.54
2593 DB + 18.10.54
2594 USSR (Lw 46, Od 52, OK 53, Od 56, i 56 Kohlenind)

52 2595 DB + 18.10.54
2596 USSR Be 1042.193 (Wysokoje 93) M
2597 CSD 555.065, 555.3065 + 68
2598 DB + 18.10.54
2599 CCCPc
52 2600 PKP *Ty2-316*
2601 PKP *Ty2-317* + *12.04.91*
2602 PKP *Ty2-318* + *13.09.88*
2603 USSR (Li 49, Ba 53, Li 56) + 01.07.82
2604 USSR (Lw 46 , Ki 48) i 51
2605 SWDE, DB (Betzdorf 50) + 18.10.54
2606 SWDE, DB (Haltingen 50) + 18.10.54
2607 USSR 1042.194 (Tschernjachowsk 85)
2608 DR (Glauchau 70) + 30.05.75
2609 PKP *Ty2-319* + *31.01.84*
52 2610 SWDE, DB (Ehrang 50) + 18.10.54
2611 USSR (SW 50) i 56
2612 USSR (Be 53), (63) PKP Ty2-1313
2613 DR (Leipzig West 70) + 25.09.80
2614 DR (Berlin Schöneweide 70) + 18.11.83
2615 USSR Be 1042.195 (Luninez 93)
2616 DR 52 8004 (Eberswalde 70) M
2617 DB + 18.10.54
2618 DB + 18.10.54
2619 ÖStB, ÖBB + 31.07.57 → Ungarn
52 2620 DR(o) KrV + 02.08.45
2621 USSR (La 56, Be 58), (62) DR (Roßlau 70) + 17.02.76
2622 USSR (Li 47, Ba 53, Li 56) + 01.07.81
2623 USSR (Okt 53, Li 58), → PKP
2624 USSR (Okt 53, Be 58), → PKP
2625 USSR, (63) MAV 520,041 + 70
2626 DR 52 8187 (Eberswalde 70) z 10.86 M
2627 DB + 18.10.54
2628 DR (Görlitz 70) + 23.04.75
2629 DB + 18.10.54
52 2630 DR v 27.07.76
2631 DR(o) KrV + 47
2632 DR (Stralsund 70) + 03.08.73
2633 USSR (Kow 46, Od 53, OK 53, i 56 Kohle) + 01.06.64
2634 USSR (Od 48, OK 53, Od 56, i 57 Kohle)
2635 USSR (Kln 54, Be 58),(63) PKP Ty2-1326
2636 DR (Berlin Schöneweide 70) + 16.05.77
2637 USSR, (64) JZ 33-288
2638 DB + 18.10.54
2639 USSR 1042.196 (Daugavpils 85) + 85
52 2640 ÖStB, ÖBB + 28.10.54
2641 PKP Ty2-339 (Nowe Skalmierzyce 91) + 18.08.89
2642 DR + 30.04.51
2643 CCCPö → CFR 150.1120 ?
2644 DR 52 8006 (Eberswalde 70) z 12.91 M
2645 PKP Ty2-1123 (Herby 94)
2646 DR (Falkenberg 70) + 24.10.75
2647 FS, (48) DB, (51) JDZ 33-260
2648 DR 52 8147 (Altenburg 70) z 06.88 M
2649 DR 52 8089 (Frankfurt/Oder 70) M
52 2650 DR 52 9650 (Senftenberg 70) + 25.10.74
2651 PKP *Ty2-322*
2652 DR 52 8120 (Altenburg 70) z 11.91 M
2653 PKP *Ty2-323*
2654 DR 52 8002 (Hoyerswerda 70) + 14.08.90
2655 USSR (OK 53, Od 56 Li 59), (63) PKP
2656 DR 52 8183 (Stendal 70) z 07.87 M
2657
2658 DR (Templin 70) + 28.09.79
2659 PKP *Ty2-324* + *23.02.90*
52 2660 CCCPd
2661 DR z 17.01.73 v 17.12.76
2662 DR (Frankfurt/Oder 70) + 19.05.76
2663 USSR (Be 47, W 51, MK 53, Kln 58) i 59
2664 PKP *Ty2-924*
2665 JDZ 33-075
2666 JDZ 33-076
2667 JDZ 33-077
2668 USSR
2669 USSR (Kln 51, Lw 60), (64) JZ

52 2670 JDZ 33-078
2671 JDZ 33-079
2672 BDZ 15.14
2673 BDZ 15.15
2674 BDZ 15.16 (Asenowo 90)
2675 DR 52 8191 (Eberswalde 70) z 08.86 M
2676 DR (Salzwedel 70) + 17.02.76
2677 DR (Dresden 70) + 24.03.77
2678 DR (Dresden 70) + 20.07.78
2679 DB + 18.10.54
52 2680 PKP *Ty2-925* + *20.02.88*
2681 DR 52 9681 (Senftenberg 70) + 23.04.75
2682 USSR (63) PKP Ty2-1363 + 10.04.87
2683 DR (Jüterbog 70) + 20.08.76
2684 USSR (i) (Lesnaja 64)
2685 USSR (Kow 46, Od 53, OK 53, i 56 Kohlenindustrie)
2686 DR (Elsterwerda 70) + 02.06.76
2687 DR (Pasewalk 70) + 15.03.73
2688 DR + 04.01.78
2689 DB + 18.10.54
52 2690 DR (Eberswalde 70) + 03.08.73
2691 DR + 20.04.77
2692 DR (Wustermark 70) + 14.11.79
2693 DR + 17.03.70
2694 PKP *Ty2-926* + *17.03.90*
2695 USSR
2696 DR 52 8151 (Bautzen 70) + 14.08.90
2697 DB + 18.10.54
2698 USSR 1042.199 (Zheleznodorozhnyj 93)
2699 JDZ 33-080
52 2700 DR + 16.05.77
2701 DR 52 8097 (Berlin Schöneweide 70) Dsp 06.88 M
2702 USSR (Kow 46, Lw 53, Li 57) + 01.03.76
2703 JDZ 33-081
2704 JDZ 33-082
2705 DR (Halle 70) + 12.07.76
2706 DR 52 8023 (Halle-G 70) z 09.87 M
2707 USSR 1042.200 (Radviliskis 85) + 89
2708 DB + 18.10.54
2709 DR (Görlitz 70) + 26.08.76
52 2710 PKP Ty2-927 (Koscierzyna 90)
2711 DR (Neustrelitz 70) + 15.03.73
2712 DR + 24.02.70
2713 DR(o) KrV + 46
2714 PKP *Ty2-928* + *18.06.86*
2715 USSR, (63) CSD 555.0270, 555.3270 + 72
2716 USSR, (63) MAV 520,042 + 04.79
2717 USSR 1042.201 (Tschernjachowsk 85) + 89
2718 DR 52 9718 (Senftenberg 70) + 02.01.80
2719 USSR Be 1042.202 (Berestowiza 93)
52 2720 DR 52 8039 (Altenburg 70) M
2721 DR (Cottbus 70) + 26.08.76
2722 PKP *Ty2-929* + *20.05.86*
2723 DR (Frankfurt/Oder 70) + 11.06.85
2724 NSB 2724 + 01. 58
2725 USSR 1042.203 (Tschernjachowsk 85)
2726 USSR, (63) PKP Ty2-1314 (Wolsztyn 83)
2727 USSR (Ungeny 78) + 86 → Vietnam x 94 ?
2728 BDZ 15.17 (Asenowo 92)
2729 BDZ 15.18 (Cervan Brjag 90)
52 2730 PKP Ty2-930 (Wegliniec 91) + 08.06.88
2731 JDZ 33-083
2732 USSR (Ba 53, La 56, Mo ..), (64) BDZ 15.249
2733 PKP *Ty2-1184*
2734 USSR Be 1042.205 (Wysokoje 93) x 96
2735 USSR (Okt 53, Li 59), (62) DR (Kamenz 70)+ 12.07.76
2736 DB, (51) JDZ 33-234
2737 USSR (Li 47, Ba 53, Es 56) + 01.12.77
2738 USSR Uk 1042.206 (Rawa Russkaja 92)
2739 CSD 555.0136, 555.3136 + 68
52 2740 USSR (SW 48, i 56) + 01.12.72
2741 DR (Jüterbog 69) + 30.09.71
2742 USSR 1042.207 (Mamonowo 93)
2743 DR (Görlitz 70) + 21.02.77
2744 CSD 555.080 + 64

52 2745	DR (Pasewalk 70)	+ 19.12.74
2746	DR	+ 12.50
2747	USSR	i 63
2748	DB	+ 18.10.54
2749	PKP *Ty2-931*	*+ 25.05.82*
52 2750	USSR (Kow 45, ... , OK 53, Od 56, La 58)	+ 01.07.71
2751	USSR (Li 47, Ba 53, La 56, Lw 61)	
	(62) DR (Seddin 70)	z 09.86 M
2752	PKP *Ty2-932*	*+ 14.06.89*
2753	DB (Kesselexplosion 47)	+ 47
2754	PKP *Ty2-933*	*+ 23.10.92*
2755	PKP *Ty2-934*	*+ 04.11.89*
2756	PKP *Ty2-325*	*+ 30.10.74*
2757	PKP Ty2-76 (Koscierzyna 90)	+ 25.09.92
2758	DR (Berlin Ostbahnhof 70)	+ 03.04.78
2759	DB	+ 18.10.54
52 2760	USSR (Be47,Li 59),(62)DR (Angermünde 70)	+ 04.08.69
2761	DB	+ 18.10.54
2762	PKP *Ty2-936*	*+ 27.08.74*
2763	DR 52 8179 (Stendal 70)	+ 19.08.88
2764	PKP *Ty2-937*	*+ 06.06.80*
2765	USSR (La 47, Ba 53, La 56)	+ 01.08.75
2766	USSR (Lw 46, BL47, Be 53, Li 59), (64) JZ 33-317	
2767	USSR (Li 48, Ba 53, Li 56), (63) PKP Ty2-1391	+ 91
2768	USSR Uk (Wjerbka 89)	
2769	PKP *Ty2-938*	*+ 25.10.74*
52 2770	NSB 2770	+ 04.70 M
2771	CSD 555.0165, 555.3165	+ 72
2772	PKP Ty2-939 (Przeworsk 89)	+ 05.10.90
2773	USSR 1042.209 (Radviliskis 85)	+ 90
2774	DR (Berlin Pankow 69)	+ 30.09.71
2775	PKP *Ty2-1174* (?)	
2776	PKP *Ty2-940*	*+ 12.09.88*
2777	PKP *Ty2-941*	*+ 01.09.84*
2778	PKP *Ty2-942*	*+ 08.10.86*
2779	DB	+ 18.10.54
52 2780	PKP *Ty2-943*	*+ 26.11.85*
2781	CSD 555.082	+ 69
2782	DB	+ 18.10.54
2783	DR (Zittau 70)	+ 06.06.77
2784	PKP *Ty2-944*	*+ 21.02.84*
2785	DR (Elsterwerda 70)	+ 06.09.75
2786	PKP Ty2-976 (Bialystok 93)	+ 15.01.88
2787	PKP *Ty2-946*	*+ 14.08.87*
2788	PKP *Ty2-947*	
2789	USSR 1042.210 (Riga 85)	+ 88
52 2790	USSR (Ungeny 78)	
2791	USSR Uk 1042.212 (Wjerbka 89)	
2792	PKP *Ty2-948*	*+ 18.02.71*
2793	(Rahmen → BDZ 15.187, Sindel 90)	
2794	PKP *Ty2-949*	
2795	JDZ 33-084	
2796	USSR Uk 1042.213 (Rawa Russkaja 92)	
2797	JDZ 33-085	
2798	CCCPö	
2799	JDZ 33-086	
52 2800	BDZ 15.19 (Jahinowo 92)	
2801	BDZ 15.20 (Kalojanowetz 92)	
2802	JDZ 33-087	M
2803	BDZ 15.21 (Bodrowo 92)	M
2804	USSR 1042.214 (Radviliskis 85)	+ 89
2805	USSR, (63) MAV 520,029	+ 01.78
2806	USSR Uk (Wjerbka 89)	
2807	PKP *Ty2-950*	*+ 14.01.87*
2808	USSR, (64) BDZ 15.255	
2809	USSR (Daugavpils 56, Ungeny 78)	
52 2810	USSR, (64) BDZ 15.229 ?	
2811	USSR 1042.217 (Mamonowo 93)	
2812	USSR Be 1042.218 (Wysokoje 93)	
2813	ÖStB, ÖBB (St Valentin 63)	+ 17.02.75
2814	USSR (Li 49, Ba 53, La 56, i 56 Kohlenindustrie)	
2815	PKP *Ty2-951*	*+ 15.11.87*
2816	PKP *Ty2-952*	*+ 15.01.86*
2817	PKP Ty2-953 (Chabowka 91)	M
2818	DR 52 9818	+ 10.04.79
2819	USSR 1042.614 (Daugavpils 85)	+ 87
52 2820	CSD 555.0104, 555.3104	+ 70
2821	USSR → PKP Ty2-1258 (Elk 1990)	+ 15.03.92
2822	USSR 1042.219 (Zheleznodorozhnyj 93)	
2823	DB	+ 18.10.54
2824	DB	+ 18.10.54
2825	DR (Eberswalde 70)	+ 17.03.76
2826	PKP Ty2-954 (Andrzejowka 91)	+ 16.09.91
2827	PKP *Ty2-955*	*+ 20.11.85*
2828	PKP *Ty2-956*	*+ 12.09.89*
2829	PKP	
52 2830	PKP *Ty2-958*	*+ 24.08.88*
2831	PKP *Ty2-959*	*+ 04.11.89*
2832	PKP *Ty2-960*	*+ 07.06.74*
2833	USSR (Kow 45, Lw 53), (64) BDZ 15.224	
2834	USSR (Lw 53), (63) MAV 520,066	+ 05.80
2835	PKP *Ty2-961*	*+ 15.05.90*
2836	USSR (i)	
2837	DB	+ 18.10.54
2838	PKP *Ty2-962*	*+ 11.03.92*
2839	USSR, (62) CSD 555.0233, 555.3233	+ 68
52 2840	USSR (Ki 48, OK 53, Mo 56), (64) BDZ 15.233	
2841	CCCPd	
2842	PKP *Ty2-963*	*+ 02.06.87*
2843	NSB 2843	+ 04.70
2844	USSR (OK 53, Ba 55, La 56)	+ 01.07.73
2845	DR (Eilenburg 70)	+ 20.04.77
2846	USSR (W 51, MK 53, Li 58)	i 62
2847	PKP *Ty2-1188*	
2848	PKP *Ty2-964*	
2849	USSR (SW 48, Es 56)	i 57
52 2850	DR (Berlin Schöneweide 70)	+ 22.11.78
2851	DB, (51) JDZ 33-239	
2852	USSR (Li 50 , Ba 53, Es 56, i 71 Matujzos)	
2853	PKP *Ty2-965*	*+ 30.04.87*
2854	CSD	
2855	DB	+ 18.10.54
2856	PKP *Ty2-966*	
2857	USSR, (62) CSD 555.0235, 555.3235	+ 71
2858	DR 52 8185 (Eberswalde 70)	z 03.88
2859	USSR (La 47, Ba 53, Es 56)	+ 01.10.72
52 2860	DB	+ 18.10.54
	52 2860" USSR (Wnz 45)	+ 12.45
2861	DR	+ 10.50
2862	DB	+ 18.10.54
2863	NSB 2863	+ 24.09.63
2864	USSR 1042.615 (Tschernjachowsk 85)	+ 88
2865	PKP	
2866	PKP *Ty2-968*	*+ 02.12.90*
2867	DR (Glauchau 70)	v 24.10.73
2868	PKP *Ty2-969*	*+ 26.07.89*
2869	DB (Minden 51)	+ 14.07.58
52 2870	PKP *Ty2-970*	*+ 20.09.88*
2871	DB (Herford 51)	+ 18.10.54
2872	PKP *Ty2-971*	*+ 05.10.90*
2873	USSR (Be 46, Li 59), (63) PKP Ty2-1377	+ 02.12.90
2874	PKP *Ty2-972*	*+ 04.09.79*
2875	DB	+ 18.10.54
2876	PKP *Ty2-973*	
2877	DB (Herford 51)	+ 18.10.54
2878	USSR Uk	
2879	USSR (Ungeny 78, Ungeny 94)	
52 2880	PKP *Ty2-974*	*+ 30.04.87*
2881	USSR, (64) JZ 33-273	
2882	USSR, (63) CSD 555.0293, 555.3293	+ 71
2883	DR 52 8109 (Eberswalde 70)	M
2884	USSR, (63) MAV 520,067 + 69 → Heizlok 520-01	
2885	DR (Salzwedel 70)	+ 02.06.75
2886	DR 52 8016 (Cottbus 70)	+ 05.04.90
2887	DR 52 8190 (Eberswalde 70)	M
2888	DB (Herford 51)	+ 18.10.54
2889	DB (Herford 51)	+ 12.05.55
52 2890	DB (Herford 51)	+ 18.10.54
2891	DB (Herford 51)	+ 18.10.54
2892	DB (Herford 51)	+ 15.08.55
2893	DB (Herford 51)	+ 18.10.54

JUNG

52 3100 - 3223   2/43 - 12/43
52 3224 - 3320   1/44 - 12/44
52 3321 - 3331   12/45 - 11/46

---

52 3100	USSR (St 44) #	
3101	JDZ 33-088	
3102	CCCPc	
3103	PKP *Ty2-326*	*+ 28.12.89*
3104	USSR (Kow 45, Ki 48)	i 51
3105	ÖStB, ÖBB 152.3105, (68) GKB	+ 76
3106	USSR (Od 44)	i 45
3107	DR (Pasewalk 70)	+ 12.07.76
3108	PKP *Ty2-327*	*+ 04.04.85*
3109	ÖStB, ÖBB 152.3109 (Graz 63), (68) GKB	+ 83 M
52 3110	ÖStB, ÖBB 152.3110 (Graz 63), (68) GKB	+ 75
3111	SWDE, DB (Worms 50)	+ 18.10.54
3112	USSR, (64) BDZ 15.169	
3113	NSB 3113	+ 24.09.63
3114	PKP *Ty2-328*	
3115	USSR (BL 45, ... , Okt 60)	+ 01.09.71
3116	USSR (BL 50, Be 53)	i 73
3117	USSR (Dno 81)	
3118	PKP *Ty2-329*	*+ 28.12.89*
3119	DR (Güsten 70)	+ 02.06.75
52 3120	CSD 555.091	
3121	SWDE, DB (Worms 50)	+ 18.10.54
3122	DB	+ 18.10.54
3123	DB	+ 18.10.54
3124	SWDE, DB (Freudenstadt 50)	+ 18.10.54
3125	SNCF 150 Y 3125	+ 55
3126	SWDE, DB (Radolfzell 50)	+ 18.10.54
3127	USSR, Uk	
3128	FS, (48) DB	+ 18.10.54
3129	DB	+ 18.10.54
52 3130	DB	+ 18.10.54
3131	DB	+ 18.10.54
3132	DB	+ 18.10.54
3133	DB	+ 18.10.54
3134	USSR (SW 48, La 57, Kln 58, Okt 61)	i 64
3135	PKP *Ty2-330*	*+ 06.04.83*
3136	USSR 1042.224 (Vilnius 85)	+ 88
3137	USSR (Be 47, SW 59)	i 60
3138	PKP *Ty2-331*	
3139	PKP *Ty2-332*	*+ 21.04.76*
52 3140	DR (Berlin Schöneweide 70)	+ 25.10.74
3141	DB	+ 18.10.54
3142	USSR, (63) CSD 555.0262, 555.3262	+ 71
3143	DB	+ 18.10.54
3144	PKP *Ty2-333*	*+ 10.08.83*
3145	DR (Görlitz 70)	v 02.07.79
3146	DR	+ 11.02.53
3147	PKP *Ty2-334*	*+ 10.12.75*
3148	USSR (Wnz 44) (47)→ ?	
3149	SNCF 150 Y 3149	+ 54
52 3150	USSR (St 44)	
3151	DB	+ 18.10.54
3152	USSR (Kln 51, Okt 55)	+ 01.02.77
3153	USSR (Ki 44, Lw 47, i 71 Papierfabrik Shidatschew)	
3154	DB	+ 18.10.54
3155	DB	+ 18.10.54
3156	CCCPö	
3157	SAAR, (57) DB	+ 20.11.58
3158	CCCPc	
3159	JDZ 33-089	
52 3160	PKP *Ty2-335*	*+ 22.05.91*
3161	USSR Be 1042.225 (Grodno 93)	+ 88
3162	USSR (Tschernjachowsk 55) 1042.226	M
3163		
3164	USSR Be 1042.227 (Brusgi 93)	
52 3165	USSR (Od 44, i 45 Buntmetallindustrie)	
3166	USSR (Derbent 46, Ki 48, Od 49, OK 53, Od 56, Li 58) (62) DR 52 8159 (Brandenburg 70)	+ 15.08.88
3167	USSR (Okt 53, Lw 61), (64) JZ 33-289	
3168	USSR (Wnz 44)	+ 47
3169	DR 52 8088 (Lübbenau 70)	v 11.75
52 3170	USSR (Okt 53, La 57, Li 58), (63) PKP	
3171	USSR (MK 53, Be 55, Es 58)	i 58
3172	USSR (Wnz 44 KrV)	
3173	USSR 1042.228 (Mamonowo 93)	
3174	ÖStB, ÖBB + 25.10.61 § 985.216	
3175	USSR (Kln 53, Be 58, Mo 61), (64) BDZ 15.268	
3176	USSR 1042.229 (Mamonowo 93)	
3177	USSR, (63) MAV 520, 003	+ 05.73
3178	SWDE, DB (Friedrichshafen 50)	+ 18.10.54
	52 3178" USSR (OK 53, Od 56 Mo 59), BDZ 15.156	
3179	PKP *Ty2-336*	*+ 06.87*
52 3180	PKP *Ty2-337*	*+ 30.11.86*
3181	USSR (W 45) → BDZ ? 15.155 ?	
3182	DB	+ 18.10.54
3183	PKP *Ty2-338*	*+ 15.12.88*
3184	SWDE, DB (Friedrichshafen 50)	+ 18.10.54
3185	USSR (OK 53, Od 56, Be 59), (63) PKP Ty2-1247	+ 76
3186	DR (Frankfurt/Oder 70)	+ 22.11.78
3187	DR (Halle-G 70)	+ 03.07.79
3188	PKP	
3189	PKP *Ty2-340*	*+ 07.06.91*
52 3190	USSR (Kow 50, Kln 51, La 57, i 58 Kohlenindustrie)	
3191	PKP *Ty2-341*	
3192	DR (Frankfurt/Oder 70)	+ 22.11.78
3193	DR 52 8060 (Halle-G 70)	M
3194	DR	+ 12.12.69
3195	PKP *Ty2-342*	*+ 11.11.85*
3196	PKP *Ty2-343*	*+ 27.08.88*
3197	PKP *Ty2-344*	*+ 29.11.90*
3198	PKP *Ty2-345*	*+ 15.05.90*
3199	USSR (SW 50)	i 56
52 3200	PKP *Ty2-346*	*+ 13.09.88*
3201	USSR Be 1042.230 (Wysokoje 93)	
3202	PKP *Ty2-347*	*+ 22.02.91*
3203	USSR (Be 50, W 51, MK 53)	i 58
3204	DR (Frankfurt/Oder 70)	+ 11.06.85
3205	DR (Wustermark 70)	+ 07.05.76
3206	PKP *Ty2-348*	*+ 02.04.91*
3207	USSR (La 47, Ba 53, La 56)	i 62
3208	PKP *Ty2-349*	*+ 19.10.90*
3209	DR 52 8139 (Berlin Schöneweide 70)	z 04.87 M
52 3210	USSR (Tschernjachowsk 53, Ungeny 78)	
3211	DR 52 8175 (Stendal 70)	+ 05.91 M
3212	CCCPd	
3213	USSR (OK 53, Od 56, Li 59), (63) PKP Ty2-1395	+ 83
3214	USSR (Be 47, Li 59) → PKP Ty2-1278	+ 04.11.89
3215	USSR (SW 49, S 53, i 57 Chemische Industrie)	
3216	DR (Dresden 70)	+ 23.04.75
3217	DR (Frankfurt/Oder 70)	v 23.10.75
3218	DR 52 8131 (Brandenburg 70)	z 08.87 M
3219	DR (Bautzen 70)	+ 02.12.75
52 3220	DR	+ 23.10.68
3221	USSR (Li 45, Kow 49, Lw53, La 58, i 58 Kohlenind.)	
3222	DR 52 9222 (Senftenberg 70)	+ 04.01.78
3223	DR 52 8070 (Hoyerswerda 70)	v 29.07.81
3224	DB	+ 18.10.54
3225	DR (Görlitz 70)	+ 30.01.80
3226	DR (Eberswalde 70)	+ 14.04.76
3227	SWDE, DB (Betzdorf 50)	+ 18.10.54
3228	USSR 1042.232 (Bagrationowsk 93)	
3229	PKP *Ty2-350*	*+ 05.04.79*
52 3230	DR, (59) BDZ 15.132 (Bodrowo 92)	
3231	DR (Neustrelitz 70)	+ 09.01.76
3232	DR	+ 19.12.74
3233	DR (Bautzen 70)	+ 08.05.78
3234	JDZ 33-090 (Titov Veles 96)	
3235	JDZ 33-091	
3236	BDZ 15.22	
3237	BDZ 15.23 (Kalojanowetz 92)	
3238	JDZ 33-092	
3239	JDZ 33-093	

52 3240 BDZ 15.24 (Mezdre 90) M
3241 BDZ 15.25
3242 BDZ 15.26 (Asenowo 92)
3243 BDZ 15.27 (Kalojanowetz 92)
3244 PKP *Ty2-351*
3245 DR (Frankfurt/Oder 70) + 22.11.78
3246 USSR, (62) CSD 555.0238, 555.3238 + 69
3247 DR, (56) BDZ 15.102 (Asenowo 92)
3248 CCCPm
3249 DR (Dresden 70) + 23.04.75
52 3250 DR 52 9250 (Senftenberg 70) + 28.10.77
3251 USSR (Li 47, Ba 53, Li 56) + 01.08.76
3252 DR 52 8199 (Templin 70) z 05.88 M
3253 USSR (Be 50, Kln 54, Li 59) → PKP
3254 DR 52 8170 (Brandenburg 70) M
3255 PKP *Ty2-975* *+ 28.06.85*
3256 PKP Ty2-180 (Suwalki 90)
3257 DR 52 8026 (Frankfurt/Oder 70) Dsp 88 M
3258 USSR, (63) MAV 520, 087 + 09.78
3259 DR + 17.03.70
52 3260 DR (Bautzen 70) v 01.10.74
3261 DR (Berlin Lichtenberg 70) + 15.11.77
3262 PKP *Ty2-977* *+ 15.02.82*
3263 USSR (SW 50, Mo 61), (64) BDZ 15.152
3264 PKP *Ty2-978* *+ 31.01.88*
3265 USSR 1042.233 (Mamonowo 93)
3266 USSR 1042.234 (Tallinn 85) + 85
3267 PKP *Ty2-979* *+ 26.06.78*
3268 CSD 555.0137, 555.3137 + 71
3269 USSR Be 1042.235 (Wysokoje 93)
52 3270 USSR (OK53, Od 56, Lw61), (62) DR + 24.02.70
3271 JDZ 33-094
3272 PKP *Ty2-980* *+ 02.06.87*
3273 PKP Ty2-981 (Milkowice 91) + 29.11.90
3274 ÖStB, ÖBB (Straßhof 63) + 28.04.77
3275 USSR (OK53, Od 56, Li 59) → PKP Ty2-1283 + 86
3276 DB + 18.10.54
3277 DB + 18.10.54
3278 DB + 18.10.54
3279 FS, (48) DB, (51) JDZ 33-252
52 3280 FS, (49) DB + 18.10.54
3281 FS, (51) DB + 18.10.54
3282 FS, (49) DB + 18.10.54
3283 FS, (47) DB + 18.10.54
3284 PKP Ty2-921 (Chojnice 90) + 16.08.93
3285 DR 52 9285 (Senftenberg 70) + 14.11.79
3286 USSR, (63) PKP Ty2-1298 (Jaworzyna 84)
3287 DR (Eberswalde 70) + 14.07.75
3288 PKP *Ty2-983* *+ 04.11.89*
3289 PKP *Ty2-984* *+ 05.10.90*
52 3290 PKP Ty2-985 + 09.09.87
3291 DB + 18.10.54
3292 USSR (Be47, Lw 62), (64) JZ 33-310
3293 DR + 02.01.69
3294 DR (Cottbus 70) + 15.03.73
3295 USSR (Be 50, Kln 54, Be 58)
(62) DR (Bautzen 70) v 01.08.77
3296 USSR Be 1042.607 (Wysokoje 93)
3297 USSR, Uk (Wjerbka 89)
3298 DB + 18.10.54
3299 PKP *Ty2-986* *+ 22.08.90*
52 3300 PKP *Ty2-987* *+ 23.02.90*
3301 PKP *Ty2-988* *+ 01.08.85*
3302 DR + 30.07.75
3303 DR (Hoyerswerda 70) + 06.06.77
3304 PKP *Ty2-989* *+ 25.10.74*
3305 KrV
3306 CSD 555.0169 + 70
3307 USSR (Lw 50, Li 57) + 01.06.77
3308 PKP *Ty2-991* *+ 17.08.84*
3309 PKP *Ty2-990*
52 3310 USSR, (63) MAV 520, 010 + 85
3311 USSR (Lw 46, Od 48, OK 53, Od 56, Li 58)
(62) DR 52 8068 (Altenburg 70) M
3312 DR (Elsterwerda 70) + 09.01.76
3313 PKP Ty2-992 (Milkowice 91) + 19.06.90
3314 ÖStB, ÖBB (Graz 63) + 27.10.78 M

52 3315 ÖStB, ÖBB (St Veit / Glan 63) + 27.10.78
3316 ÖStB, ÖBB (St Veit / Glan 63) + 20.01.78 M
3317 ÖStB, ÖBB (St Veit / Glan 63) + 26.05.76
3318 DR 52 8112 (Wustermark 70) + 05.04.90
3319 ÖStB, ÖBB (Wiener Neustadt 63) + 28.09.73
52 3320 PKP (Rahmen → Ty42-148)
3321 SAAR, DB + 20.07.59
3322 SAAR, DB + 17.08.61
3323 SWDE, DB (Worms 50) + 18.10.54
3324 SWDE, DB (Worms 50) + 18.10.54
3325 SWDE, DB (Worms 50) + 18.10.54
3326 SAAR, DB + 11.05.60
3327 SWDE, DB (Worms 50) + 18.10.54
3328 SAAR, DB + 09.01.62
3329 SAAR, DB + 10.08.62
52 3330 SWDE, DB (Mainz 50) + 54
3331 SAAR, DB + 11.05.60

---

52 3350 - 52 3623

KRAUSS - MAFFEI

52 3350 - 3619 3/43 - 11/43
52 3620 9/43 mit Krauss-Wellrohrkessel
52 3621 - 3622 44 mit Krauss-Wellrohrkessel
52 3623 2/45 mit Krauss-Wellrohrkessel

---

52 3350 DR (Bautzen 70) + 20.09.72
3351 USSR (St 44)
3352 USSR (Od 44 Kx, + 51) → BDZ 15.243 ?
3353 DR + 17.01.72
3354 USSR (Len 51, Okt 53, Be 58)
(62) DR 52 8116 (Hoyerswerda 70) z 03.88 M
3355 USSR (NK 46, Ki 49, i 51) → BDZ 15.206
3356 USSR 1042.237 (Kaliningrad 93)
3357 PKP *Ty2-352* *+ 17.03.90*
3358 USSR (St 44)
3359 USSR (St 44)
52 3360 USSR (BL 50, Be 53, i 57 Kohlenindustrie)
3361 PKP *Ty2-353* *+ 16.10.91*
3362
3363 USSR (Be 53, Li 59), (62) DR (Zittau 70) + 14.04.76
3364 ÖStB, ÖBB (Graz 63) + 20.10.74
3365 FS, (48) DB, (51) JDZ
3366 PKP *Ty2-354* *+ 18.03.72*
3367 ÖStB, ÖBB 152.3367 (Wiener Neustadt 63)+ 01.02.73
3368 USSR 1042.238 (Valga 89) M
3369 JDZ + 46
52 3370 ÖStB KrV + 45
3371 USSR (Luninez 90)
3372 ÖStB, ÖBB (Knittelfeld 63) + 29.11.77
3373 DB + 18.10.54
3374 DB + 18.10.54
3375 USSR (Kln 53, Be 57, Mo 61), (64) BDZ 15.275
3376 ÖStB KrV + 30.04.48
3377 DB + 18.10.54
3378 DB, (51) JDZ
3379 DB + 18.10.54
52 3380 PKP *Ty2-355* *+ 19.08.85*
3381 DR (Görlitz 70) + 10.04.79
3382 USSR, Uk
3383 DR + 20.12.51
3384 USSR + 88
3385 USSR (Od 44-47) (Mederowo 46) + 48
52 3385" USSR, Uk 1042.242 (Rawa Russkaja 92)
3386 FS, (49) DB + 18.10.54
3387 USSR 1042.243 (Krasnowka 93)
3388 USSR
3389 USSR (Od 45, Lw48, Od 48 , OK 53, Od 56) i 57
52 3390 USSR (OK 53, Ba 54, Es 56) i 63
3391 USSR (Lw 44) PKP *Ty2-356* *+ 19.11.91*
3392 DR (Altenburg 70) + 07.08.79
3393 USSR (Lw 48) + 01.08.74
3394 CSD 555.0117, 555.3117 + 68

52 3395 DR, (59) BDZ 15.127 ?
3396
3397 USSR 1042.245 (Gusew 92) M
3398 PKP *Ty2-357* *+ 19.10.90*
3399 PKP *Ty2-358* *+ 14.07.75*
52 3400 USSR, (63) CSD 555.0282, 555.3282 + 69
3401 SWDE, DB (Mainz 50) + 18.10.54
3402 USSR (OK53, Od 56, Li 58), (63) PKP Ty2-1302 + 90
3403 DR 52 8110 (Cottbus 70) x 10.91
3404 PKP *Ty2-359* *+ 22.09.87*
3405 PKP *Ty2-360* *+ 19.11.91*
3406 PKP *Ty2-361* *+ 23.02.90*
3407 SWDE, DB (Ehrang 50) + 18.10.54
3408 USSR (Li 48, Ba 53, Es 56) i 58
3409 PKP *Ty2-362* *+ 16.06.86*
52 3410 DR (Falkenberg 70) + 08.04.86
3411 USSR (Ba 53, La 56, Mo 59), (64) BDZ 15.189
3412 JDZ 33-095
3413 JDZ 33-096
3414 USSR (W 47, Ki 49) i 51
3415 USSR (OK53, Mo 56), (64) BDZ 15.234
3416 PKP *Ty2-1181* *+ 06.05.89*
3417 DB (Simbach 50), (51) JDZ 33-253 M
3418 USSR, (62) CSD 555.0206, 555.3206 + 72
3419 USSR 1042.246 (Kaliningrad s93)
52 3420 DR 52 8098 (Halle-G 70) # M
3421 USSR (Wnz 44, Od 49, OK 53) i 55
3422 USSR (BL 47, Be 53), (63) PKP
3423 PKP *Ty2-363* *+ 26.07.89*
3424 CSD 555.092
3425 PKP *Ty2-364* *+ 28.12.87*
3426 PKP *Ty2-365* *+ 25.08.72*
3427 USSR (Od 44, i 45 Buntmetallindustrie)
3428
3429
52 3430 DB + 18.10.54
3431 DB + 18.10.54
3432 JDZ 33-097
3433 PKP *Ty2-1173*
3434 USSR 1042.247 (Tschernjachowsk 93)
3435 PKP *Ty2-366* *+ 30.07.91*
3436 JDZ 33-098
3437 NSB 3437 + 04.70
3438 PKP *Ty2-367* *+ 23.02.90*
3439 USSR (Okt 58 Mo 60), Export?
52 3440 ÖStB, ÖBB (Mürzzuschlag 63) + 22.11.72
3441 ÖStB, ÖBB + 25.01.61
3442 PKP *Ty2-368* *+ 17.03.90*
3443 USSR, (63) MAV 520,049 # + 12.79
52 3443" USSR (Od 47 Kx) + 48
3444 DB + 18.10.54
3445 USSR
3446 ÖStB, ÖBB (Knittelfeld 63) + 29.11.77
3447 JDZ KrV 33-099 ? + 46
3448 DB + 18.10.54
3449 DR (Wustermark 70) + 04.01.77
52 3450 DB, (51) JDZ 33-246
3451 CSD 555.0161 + 69
3452 USSR (Kln 51, Okt 55, Be 60), (62) PKP
3453 USSR 1042.250 (Mamonowo 92)
3454 USSR (Lw 52) + 01.04.74
3455 DR 52 8044 (Leipzig West 70) z 05.88 M
3456 PKP *Ty2-369*
3457 DR 52 9457 (Senftenberg 70) + 17.03.76
3458 USSR (Kln 51, Okt 61, Be 63) → Pi (Szczakowa 91)
3459 USSR (Od 50, OK 53, Ba 55, La 56, Li 57)
(62) DR (Neustrelitz 70) + 04.08.69
52 3460 CSD
3461 CSD
3462 USSR (OK 53, Mo 56), Export?
3463 USSR 1042.251 (Tallinn 90) + 89
3464 DR 52 8015 (Cottbus 70) M
3465 PKP
3466 DB + 18.10.54
3467 DB + 18.10.54
3468 USSR, (49) → TE 8009 (OK 53, Od56, i 56 Kohle) + 64
3469 USSR (Kln 51, La 57) i 58

52 3470 DB + 18.10.54
3471 ÖStB, ÖBB + 31.07.57 → Ungarn
3472 PKP *Ty2-371* *+ 25.08.88*
3473 DR 52 8128 (Eberswalde 70) + 24.04.87
3474 DB + 18.10.54
3475 USSR (Kow 45, BL 47, Be 53, Lw 61)
(62) DR (Wustermark 70) + 20.12.79
3476 ÖStB, ÖBB + 21.12.57 → Ungarn
3477 DR (Kamenz 70) + 09.01.76
3478 DR (Dresden 70) + 12.07.76
3479 DB + 18.10.54
52 3480 USSR (Ki 44, Lw 47, Ki 48, OK 53) i 55
3481 USSR (Ba 53, Li 56), (63) PKP Ty2-1364 + 12.11.85
3482
3483 SNCF 150 Y 3483 + 55
3484 PKP *Ty2-372* *+ 04.11.89*
3485 USSR (SW 50, S 53) + 57
3486 PKP *Ty2-373* *+ 11.02.86*
3487 USSR (Od 45, OK 53, Od 56, i 57 Kohlenindustrie)
3488 USSR (Kow 46, Wnz 52, SW 53, Es 56) i 57
3489 DR + 01.51
52 3490 USSR (Li 46, Be 47, Lw 61), (66) JZ 33-323
3491 PKP *Ty2-374* *+ 23.11.79*
3492 USSR (Li 45, SW 49, i 56 Kohlenindustrie)
3493 PKP *Ty2-375*
3494 USSR (SW 53, Od 57, Be 62) → Pi
3495 PKP *Ty2-376* *+ 07.03.85*
3496 PKP *Ty2-377*
3497 PKP *Ty2-378*
3498 DR (Berlin Schöneweide 70) + 30.01.80
3499 SWDE, DB (Mainz 50) + 18.10.54
52 3500 DB + 18.10.54
3501 DR + 23.02.76
3502 USSR, (63) CSD 555.0257 + 69
3503 ÖStB, ÖBB (Straßhof 63) + 29.11.77
3504 ÖStB, ÖBB (Straßhof 63) + 10.09.78 M
3505 USSR, Uk (Rawa Russkaja 92)
3506 CCCPm → CFR 150.1118 ?
3507 ÖStB, ÖBB (Straßhof 63) + 30.07.73
3508 USSR (BL 47, Be 53, i 57 Kohlenind) + 01.02.70
3509 ÖStB, ÖBB (Wiener Neustadt 63) + 29.11.77
52 3510 ÖStB, ÖBB + 20.11.54
3511 ÖStB, ÖBB + 20.11.54
3512 ÖStB, ÖBB (Graz 63) + 15.06.73
3513 PKP *Ty2-379* *+ 30.04.87*
3514 USSR (SW 50, Mo 60) + 01.10.77
3515 USSR (Lw 51), (64) BDZ 15.219
3516 DB + 18.10.54
3517 ÖStB, ÖBB (Selzthal 63) + 27.10.78 M
3518 ÖStB, ÖBB (Mürzzuschlag 63) + 25.04.73
3519 ÖStB, ÖBB + 20.11.54
52 3520 ÖStB, ÖBB (Graz 63) + 25.07.77 § 985.226
3521 ÖStB, ÖBB + 14.09.57 → Ungarn
3522 USSR (Len 51, Okt 53, Es 56) i 62
3523 FS, (48) DB + 49
3524 CSD 555.095 + 72
3525 USSR (Kow 46, Lw 53) + 01.04.73
3526 CSD + 70
3527 DB + 18.10.54
3528 USSR (Lw 53), (63) MAV 520, 078 + 04.77
3529 DB + 18.10.54
52 3530 CSD 555.0603, 555.3184 + 71
3531 DB + 18.10.54
3532 DB + 18.10.54
3533 DB + 18.10.54
3534 DB + 18.10.54
3535 USSR, (63) MAV 520, 030 , (76) GySEV + 84 M
3536 SNCF 150 Y 3536
3537 DB + 18.10.54
3538 DB + 18.10.54
3539 USSR, Uk 1042.253 (Rawa Russkaja 92)
52 3540 ÖStB, ÖBB (Knittelfeld 63) + 25.04.73
3541 DB + 18.10.54
3542 DB + 18.10.54
3543 PKP *Ty2-380* *+ 15.05.90*
3544 USSR, (63) CSD 555.0271, 555.3271 + 69

52 3545 USSR 1042.254 (Vilnius 85) + 88
3546 DR (Berlin Schöneweide 70) + 20.04.77
3547 USSR (Li 47, Ba 53, Li 56) + 01.07.73
3548 DR (Görlitz 70) z 19.02.87 M
3549 DB + 18.10.54
52 3550 DR 52 8031 (Frankfurt/Oder 70) x 03.92
3551 DB + 18.10.54
3552 PKP *Ty2-381* *+ 26.07.89*
3553 ÖStB, ÖBB (Straßhof 63) + 10.01.73 § 985.222
3554 USSR,(Be 50) → Pi Ty2-3554 M
3555 ÖStB, ÖBB (Selzthal 63) + 29.12.75
3556 USSR,Uk
3557 ÖStB, ÖBB (Selzthal 63) + 29.11.77
3558 USSR (49) → TE-8003 (Krasnowka 93)
3559 USSR (Ki 45, Lw 48, Od 48) i 51
52 3560 ÖStB, ÖBB (Mürzzuschlag 63) + 15.07.70
3561 JDZ 33-100
3562 DB (München Ost 50) + 18.10.54
3563 JDZ 33-101
3564 USSR Rahmen → TE-8005 ?
3565 USSR,(62) DR (Brandenburg 70) + 15.11.77
3566 PKP *Ty2-382* *+ 06.04.82*
3567 USSR (Be 53, Li 59) → PKP
3568 DR 52 8115 (Eberswalde 70) z 05.88 M
3569 USSR (La 50, Ba 53, La 56, i 64 Riga) + 73
52 3570 DB + 18.10.54
3571 CCCPd PKP *Ty2-383* *+ 13.11.75*
3572 PKP *Ty2-384* *+ 21.05.76*
3573 DR (Elsterwerda 70) + 24.10.75
3574 DR 52 8168 (Stendal 70) M
3575 USSR (Lw 44) (48) → ?
3576 ÖStB, ÖBB + 20.11.54
3577 USSR (Ki 48, Wnz 51, SW 53) i 56
3578 PKP *Ty2-385* *+ 11.04.92*
3579 PKP *Ty2-386* *+ 20.09.89*
52 3580 DR(o) + 08.47
3581 USSR (Li 47, Ba 53, Li 56) + 01.07.80
3582 USSR, (64) BDZ 15.239 (Kalojanowetz 92)
3583 DB + 18.10.54
3584 USSR (Li 47, Ba 53, Li 56) i 64
3585 DB + 18.10.54
3586 DB + 18.10.54
3587 USSR (SW 50, S 53) + 57
3588 DR (Cottbus 70) + 23.04.75
3589 PKP *Ty2-387* *+ 02.04.76*
52 3590 USSR, (62) DR 52 8188 (Eberswalde 70) + 06.02.78
3591 SWDE, DB (Betzdorf 50) + 18.10.54
3592 USSR(i)
3593 FS, (49) DB + 18.10.54
3594 DR 52 9594 (Senftenberg 70) + 24.03.77
3595 SNCF 150 Y 3595 + 54
3596 DB, (51) JDZ 33-262
3597 DB + 13.12.50
3598 DB + 18.10.54
3599 DR(w) + 20.09.48
52 3600 DB + 18.10.54
3601 DB + 18.10.54
3602 ÖStB, ÖBB + 28.10.58 § 985.209
3603 DR + 12.12.69
3604 PKP *Ty2-388*
3605 USSR (Be 47), (62) PKP Ty2-1229 + 10.12.75
3606 NSB 3606 + 04.70
3607 USSR, Uk 1042.256 (Rawa Russkaja 92)
3608 PKP *Ty2-389* *+ 12.02.86*
3609
52 3610 USSR, (63) MAV 520, 068 + 11.72
3611 USSR (Be 50, Be 58) + 01.04.68
3612 USSR (Königsberg Wrk 46) + 48
3613 ÖStB KrV + 47
3614 ÖStB, ÖBB + 20.11.54
3615 ÖStB, ÖBB (Straßhof 63) + 11.10.76 § 985.225
3616 ÖStB, ÖBB (Selzthal 63) + 29.11.77
3617 ÖStB, ÖBB + 20.11.54
3618 ÖStB KrV + 48
3619 ÖStB, ÖBB (Wels 63) + 16.06.75
52 3620 DB + 11.52
3621 DR(w)
3622 DB + 11.52
3623 DB + 11.52
(3624 Nicht fertiggestellt)

52 3625 - 52 4049 Bestellt bei Krauss-Maffei

	Gebaut von:	Anlieferung:
52 3625 - 52 3631	WLF	10/43 - 11/43
52 3632 - 52 3634	DWM	11/43
52 3635 - 52 3637	MBA	2/44
52 3638 - 52 3639	Esslingen	10/43
52 3640 - 52 3641	Krenau	10/43 - 11/43
52 3642 - 52 3655	WLF	11/43
52 3656 - 52 3661	DWM	11/43 - 12/43
52 3662 - 52 3667	MBA	2/44
52 3668 - 52 3670	Esslingen	12/43
52 3671 - 52 3674	Krenau	11/43
52 3675 - 52 3688	WLF	12/43 - 1/44
52 3689 - 52 3694	DWM	12/43 - 2/44
52 3695 - 52 3700	MBA	3/44
52 3701 - 52 3704	Esslingen	12/43 - 1/44
52 3705 - 52 3708	Krenau	12/43 - 1/44
52 3709 - 52 3722	WLF	1/44 - 2/44
52 3723 - 52 3728	DWM	2/44 - 3/44
52 3729 - 52 3731	MBA	3/44
52 3732 - 52 3735	Esslingen	1/44 - 2/44
52 3736 - 52 3739	Krenau	1/44 - 2/44
52 3740 - 52 3753	WLF	2/44 - 4/44
52 3754 - 52 3759	DWM	3/44 - 4/44
52 3760 - 52 3761	Esslingen	2/44
52 3762 - 52 3763	(Esslingen)	52 1523 & 52 1525
52 3764 - 52 3767	Krenau	2/44 - 3/44
52 3769 - 52 3781	WLF	4 /44 - 5/44
52 3782 - 52 3787	DWM	5/44 - 6/44
52 3788 - 52 3796	MBA	4/44 - 5/44
52 3797 - 52 3800	Esslingen	4/44
52 3801 - 52 3804	Krenau	4/44 - 5/44
52 3805 - 52 3818	WLF	5/44
52 3819 - 52 3824	DWM	6/44 - 7/44
52 3825 - 52 3833	MBA	5/44
52 3834 - 52 3837	Esslingen	4/44 - 6/44
52 3838 - 52 3841	Krenau	5/44 - 8/44
52 3842 - 52 3855	WLF	5/44
52 3856 - 52 3861	DWM	7/44 - 8/44
52 3862 - 52 3870	MBA	5/44 - 6/44
52 3871 - 52 3874	Esslingen	6/44
52 3875 - 52 3878	Krenau	6/44
52 3879	WLF	5/44
52 3880 - 52 3887	DWM	8/44
52 3892 - 52 3900	MBA	6/44
52 3901 - 52 3906	Krenau	6/44 - 7/44
52 3913 - 52 3918	MBA	6/44 - 7/44
52 3919 - 52 3925	Krenau	7/44
52 3938 - 52 3943	MBA	8/44
52 3950 - 52 3961	Krenau	7/44 - 9/44
52 4045 - 52 4049	WLF	6/44

52 3625 DR 52 8180 (Brandenburg 70) z 08.87 M
3626
3627 DR(o) + 05.02.49
3628 DR + 17.02.76
3629 ÖStB, ÖBB (Straßhof 63) + 25.04.73
52 3630 DR (Berlin Schöneweide 70) + 07.08.79
3631 DB + 18.10.54
3632 PKP
3633 DR 52 8054 (Falkenberg 70) + 06.03.91
3634
3635 DR (59) BDZ 15.128 ?
3636 ÖStB, ÖBB + 25.01.61 § 985.214
3637 DR (Altenburg 70) + 30.01.80
3638 USSR (Od 50, OK 53) i 55
3639 USSR (Be 50), (63) PKP Ty2-1249
52 3640 PKP Ty2-18 (Olesnica 91) + 11.04.92
3641 PKP *Ty2-392* *+ 15.01.86*
3642 USSR (Be 58), (62) PKP Ty2-1223 + 06.04.82
3643 USSR 1042.257 (Krasnowka 93)
3644 USSR 1042.258 (Kaliningrad 92) M
3645 USSR (Od 50, OK 53, Od 56, Li 58) → i
3646 USSR (Kow 46, Lw 53, La 57) i 75
3647 DB + 18.10.54
3648 DB + 21.04.52
3649 DB + 18.10.54
52 3650 DB + 18.10.54
3651 USSR (Len 51, Okt 53, Li 59), (63) PKP Ty2-1273 + 82
3652 DB + 18.10.54
3653 DB + 18.10.54
3654 USSR (Kln 51, Li 59), (64) Jugoslawien 33-505
3655 DB + 18.10.54
3656 PKP *Ty2-393* *+ 11.04.92*
3657 SWDE,DB (Friedrichshafen 50) + 18.10.54
3658 PKP Ty2-394 (Stalowa Wola 93)
3659 PKP *Ty2-395* *+ 20.05.87*
52 3660 ÖStB, ÖBB (Graz 63) + 15.07.70
3661 DB + 18.10.54
3662 PKP *Ty2-994* *+ 20.05.86*
3663 USSR (Okt 53, i 57 Kohlenindustrie)
3664 ÖStB, ÖBB + 25.04.56
3665 USSR 1042.259 (Mamonowo 93)
3666 DB + 54
3667 USSR 1042.260 (Radviliskis 85) + 89
3668 DB + 18.10.54
3669 JDZ 33-102
52 3670 USSR, (62) CSD 555.0204 555.3204 + 68
3671 USSR (OK 53, Mo 56), BDZ 15.168 ?
3672 USSR (Len 51, Okt 53, i 57 Kohlenindustrie)
3673 DR 52 8103 (Leipzig West 70) + 06.03.91
3674 DR(w), (45) USA
3675 USSR Uk (Wjerbka 89)
3676 USSR (Be 47, SW 48, i 56 Buntmetallindustrie)
3677 PKP *Ty2-396* *+ 15.10.65*
3678
3679 USSR (W 48, Kln 51, Li 59), → PKP Ty2-1291
52 3680 USSR (Od 48, OK 53, Od 56, i 57 Kohlenindustrie)
3681 ÖStB, ÖBB (Graz 63) + 16.08.76
3682 ÖStB, ÖBB + 27.12.62
3683 ÖStB, ÖBB (Straßhof 63) + 01.02.73
3684 DB + 18.10.54
3685 JDZ 33-103
3686 ÖStB, ÖBB + 07.04.53
3687 ÖStB, ÖBB + 05.02.52
3688 ÖStB, ÖBB (Knittelfeld 63) + 01.02.73
3689 USSR (SW 48, i 56 Buntmetallindustrie)
52 3690 PKP *Ty2-397* *+ 29.11.90*
3691 PKP *Ty2-398* *+ 13.09.88*
3692 DB + 18.10.54
3693 USSR Uk 1042.262 (Rawa Russkaja 92)
3694 DB + 18.10.54
3695 SWDE, DB (Singen 50) + 18.10.54
3696 USSR, (62) CSD 555.0229 555.3229 + 68
3697 SNCF 150 Y 3697
3698 PKP *Ty2-995* *+ 04.11.89*
3699 DR 52 8083 (Cottbus 70) z 88 M

52 3700 PKP *Ty2-996* *+ 06.06.84*
3701 PKP *Ty2-399* *+ 15.09.89*
3702 JDZ 33-189
3703 USSR (SW 50, i 56 Buntmetallindustrie)
3704 PKP *Ty2-997* *+ 12.09.89*
3705 DR (Wustermark 70) + 14.04.76
3706 PKP *Ty2-1182* *+ 31.05.83*
3707 DB + 18.10.54
3708 USSR
3709 CSD 555.0159, 555.3159 + 72
52 3710 USSR (Be 58), Uk
3711 PKP *Ty2-998* *+ 15.12.87*
3712 USSR Be 1042.265 (Wysokoje 93)
3713 PKP *Ty2-999* *+ 28.08.78*
3714 PKP *Ty2-400* *+ 20.06.81*
3715 PKP Ty2-1000 (Jaworzyna 83) + 14.08.84
3716 SWDE, DB (Kirn 50) + 18.10.54
3717 DR (Berlin Schöneweide 70) + 30.01.80
3718 USSR Uk 1042.266 (Rawa Russkaja 92)
3719 CCCPc
52 3720 USSR 1042.267 (Bagrationowsk 93)
3721 DR 52 8049 (Frankfurt/Oder 70) x 11.92
3722 DR 52 8184 (Brandenburg 70) M
3723 DB + 18.10.54
3724 USSR (Kln 51, Be 60), (63) PKP Ty2-1256+ 20.01.73
3725 PKP *Ty2-1001* *+ 31.01.88*
3726 DR (Frankfurt/Oder 70) + 20.02.75
3727 USSR (Be 58), Uk
3728 PKP *Ty2-1002* *+ 19.07.79*
3729 PKP *Ty2-1003* *+ 20.12.89*
52 3730 PKP *Ty2-1004* *+ 18.04.72*
3731 PKP *Ty2-1005* *+ 07.06.91*
3732 PKP *Ty2-1006* *+ 10.04.92*
3733 DB + 18.10.54
3734 DR 52 8013 (Cottbus 70) M
3735 USSR, (63) MAV 520,080 + 08.73
3736 USSR (Okt 55, Be 58), (63) PKP Ty2-1346 + 12.90
3737 USSR (Kln 52, Okt 55, Mo60), BDZ 15.143
3738 DB, (51) JDZ
3739 PKP Ty2-1007 (Malaszewicze 96)
52 3740 DB + 18.10.54
3741 DR (Görlitz 70) + 17.02.76
3742 DR (Glauchau 70) + 23.04.75
3743 CCCPc
3744 USSR 1042.269 (Krasnowka 93)
3745 PKP *Ty2-1008* *+ 01.08.85*
3746 USSR (Okt 53) + 01.12.77
3747 USSR (Be 50, Sw 59, Lw 61)
(62) DR (Frankfurt/Oder 70) + 25.08.78
3748 PKP *Ty2-1009* *+ 11.04.84*
3749 USSR 1042.270 (Riga 85) + 87
52 3750 PKP *Ty2-1010*
3751 USSR 1042.271 (Kaliningrad 93)
3752 USSR (Be 50, Li 57) + 01.10.73
3753 USSR (Okt 53, Be 60), (63) PKP Ty2-1239 + 02.90
3754 USSR (Lw 48), (63) CSSR
3755
3756 PKP *Ty2-1011* *+ 08.11.74*
3757 DB + 18.10.54
3758 NSB 3758 + 09.63
3759 USSR (Li 56, Be 59), (63) PKP Ty2-1355 + 08.83
52 3760 CCCPd
3761 DR (Zwickau 70) + 17.08.81
3762 USSR, (63) MAV 520,088 + 05.77
3763 DB + 18.10.54
3764 USSR (BL 50, Be 53) + 01.09.68
3765 USSR 1042.272 (Zheleznodorozhnyj 93)
3766 DB + 18.10.54
3767 USSR , (63) MAV 520,069 # + 04.79
3768 wurde als 52 3879 abgeliefert.
3769 JDZ 33-214
52 3770 PKP Ty2-1012 (Milkowice 91) + 15.11.91
3771 PKP *Ty2-1013* *+ 20.06.90*
3772 PKP *Ty2-1014* *+ 17.03.90*
3773 CCCPd
3774 BDZ 15.28 (Asenowo 92)

52 3775 BDZ 15.29 (Bodrowo 92)
3776 DB + 18.10.54
3777 BDZ 15.30 (Asenowo 90)
3778 PKP *Ty2-1015* *+ 12.02.86*
3779 USSR (Okt 53, Be 58), (63) PKP
52 3780 USSR (OK 53, Od 56, Mo58), BDZ 15.144
3781 USSR (Len 51, Okt 53, Es 57) i 58
3782 PKP *Ty2-1016* *+ 09.03.85*
3783 PKP *Ty2-1017*
3784 DR (Salzwedel 70) + 20.04.77
3785 USSR 1042.273 (Riga 85) + 90
3786 USSR (Okt 53, Li 62), (63) PKP Ty2-1358 + 07.03.85
3787 DB + 13.12.50
3788 DR (Frankfurt/Oder 70) + 23.01.79
3789 USSR (Be 53, i 57 Kohlenind, Do 61, i 62 Kohlenind)
52 3790 USSR (Okt 53, Be 58), (62) DR (Görlitz 70) + 01.09.80
3791 USSR (Ba 53, La 56, Mo 60), BDZ 15.145 (Mezdra 90)
3792 PKP
3793 CCCPd → PKP *Ty2-1019* *+ 10.09.88*
3794 DR 52 9794 (Senftenberg 70) + 19.05.76
3795 PKP *Ty2-401* *+ 20.11.87*
3796 DR (Berlin Schöneweide 70) + 13.10.80
3797 USSR (Lw 51), (64) BDZ 15.207 (Kalojanowetz 92)
3798 USSR (SW 50, Be 58), (62) PKP Ty2-1209 + 05.73
3799 USSR Uk (Wjerbka 89)
52 3800 USSR (Lw 49, Od 50, OK 53, Od 56, i 56 Kohlenind)
3801 USSR (Kln 53, Li 58), (63) PKP
3802 PKP *Ty2-1020* *+ 04.11.89*
3803 USSR, (62) CSD 555.0240 555.3240 + 70
3804 DR (Zittau 70) + 14.07.75
3805 BDZ 15.31
3806 BDZ 15.32
3807 BDZ 15.33 (Bodrowo 92)
3808 BDZ 15.34 (Kalojanowetz 92)
3809 USSR + 88
52 3810 ÖStB, ÖBB (Wien Ost 63) + 01.02.73
3811 SNCF 150 Y 3811
3812 ÖStB, ÖBB (St. Pölten 63) + 28.04.76
3813 PKP Ty2-432 (Elk 90)
3814 JDZ 33-104
3815 USSR, (63) MAV 520,050 (10.63) GySEV + 03.80
3816 ÖStB, ÖBB (Graz 63) + 04.77 M
3817 ÖStB, ÖBB (Straßhof 63) + 16.08.76
3818 ÖStB, ÖBB (Wien Ost 63) + 05.05.71
3819
52 3820 USSR (Be 58), (62) DR (Görlitz 70) + 22.02.79
3821 USSR (Be 50, Lw 62), (64) JZ 33-303
3822 CSD 555.0105
3823 USSR (OK 53, Mo56), (63) MAV 520,051 + 11.72
3824 PKP *Ty2-1022* *+ 05.10.90*
3825 USSR (Od 50, OK 53, Od 56, i 57 Kohlenind) + 62
3826 USSR Be 1042.276 (Brusgi 93)
3827 ÖStB, ÖBB + 14.09.57 → Ungarn
3828 PKP *Ty2-1023* *+ 07.03.87*
3829 USSR (Kln 59, Mo61), (64) JZ 33-297
52 3830 USSR (Okt 53, Mo60), BDZ 15.146
3831 USSR(i) Lesnaja 70
3832 NSB 3832 + 24.09.63
3833 CCCPc
3834 DR (Neubrandenburg 70) + 29.07.76
3835 USSR (Okt 53, Mo61), (64) BDZ 15.241 (Bodrowo 92)
3836 CCCPc
3837 USSR Be 1042.277 (Brest Wostok 93)
3838 USSR 1042.278 (Tschernjachowsk 85)
3839 DR 52 8149 (Elsterwerda 70) M
52 3840 PKP *Ty2-1024* *+ 12.10.86*
3841 USSR (La 47, Ba 53, La 56) + 01.01.73
3842 USSR (Ostaschkow 88) + 06.05.88
3843 DR + 20.12.51
3844 BDZ 15.35
3845 BDZ 15.36
3846 BDZ 15.37
3847 BDZ 15.38 (Mezdra 90)
3848 JDZ 33-219
3849 BDZ 15.39 (Bodrowo 92)

52 3850 BDZ 15.40 (Kalojanowetz 92)
3851 BDZ 15.41 (Kalojanowetz 92)
3852 BDZ 15.42
3853 JDZ 33-225
3854 BDZ 15.43 (Bodrowo 92)
3855 BDZ 15.44 (Bodrowo 92)
3856 JDZ 33-190
3857 USSR (Be 47, Li 58, i 68 Gusew)
3858 USSR (Be 50) → PKP Ty2-1282 + 05.04.91
3859
52 3860 USSR (Be 47, Okt 58, Mo 60), (64) JZ 33-298
3861 USSR 1042.280 (Kaliningrad 85) + 89
3862 DB + 18.10.54
3863 DB + 18.10.54
3864 USSR (Li 50, Ba 53, Li 56) + 01.04.72
3865 DB + 18.10.54
3866 USSR (Lw 53, La 57, Li 59), (63) PKP
3867 PKP *Ty2-402* *+ 18.08.89*
3868 USSR 1042.281 (Zheleznodorozhnyj 93)
3869 USSR 1042.282 (Zheleznodorozhnyj 93)
52 3870 PKP Ty2-1025 (Chelm 92) + 05.10.90
3871 DR 52 8176 (Brandenburg 70) Dsp 88, M
3872 USSR (Ba 53, La 56, Mo ..), (64) BDZ
3873 PKP *Ty2-1026*
3874 PKP *Ty2-1027* *+ 25.10.85*
3875 PKP *Ty2-1028*
3876 USSR (Be 47, Lw 61), (64) JZ
3877 DR(o) + 47
3878 DB + 18.10.54
3879 ÖStB, ÖBB (Graz 63) + 10.78 M
52 3880 PKP ? *Ty2-1029* ?
3881 DB + 18.10.54
3882 PKP *Ty2-1194* *+ 18.10.81*
3883 USSR (Li 45, Be 48, La 57, Be 58), (63) Ty2-1351 + 83
3884 USSR (OK 53, Od 56, La 57, Mo 60), BDZ 15.147
3885 PKP *Ty2-1030* *+ 09.12.85*
3886 PKP *Ty2-403* *+ 29.11.90*
3887 USSR (Kow 45, Lw 53), (63) CSSR

---

52 3892 PKP *Ty2-1031* *+ 17.03.90*
3893 USSR (Li 45, La 47, Ba 53, La 56) + 01.10.70
3894 USSR (Li 45, Be 47, Li 58, i 58, Do 61) i 61
3895 USSR (Od 48, OK 53, i 56, Do 61) i 61
3896
3897 USSR, (63) MAV 520,020 (64) GySEV + 10.79
3898 USSR (Li 45)
3899 USSR (Li 45, Be 47, W 51, Kln 53, Be 57, Li 58) i 58
52 3900 USSR, (62) CSD 555.0216 555.3216 + 68
3901 USSR (Li 45, La 48, Ba 53, La 56, i 56 Kohlenind)
3902 PKP *Ty2-1032* *+ 16.04.74*
3903 PKP *Ty2-1033* *+ 04.04.85*
3904 PKP *Ty2-1034* *+ 22.05.91*
3905 USSR (Li 47, Ba 53, Es 56) + 01.03.72
3906 DB + 18.10.54

---

52 3913 SWDE , DB (Friedrichshafen 50) + 18.10.54
3914 PKP
3915 USSR Be 1042.283 (Wysokoje 93) M
3916 DR 52 8017 (Cottbus 70) M
3917 SWDE, DB (Friedrichshafen 50) + 18.10.54
3918 CCCPc
3919 USSR (Okt 53, Li 58), (63) PKP Ty2-1264 + 10.06.68
52 3920 CSD 555.0116, (50) 555.0606 ...
... (64) 555.0183, 555.3183 + 68
3921 PKP *Ty2-1036* *+ 20.05.87*
3922 PKP *Ty2-1037* *+ 13.03.87*
3923 USSR 1042.284 (Krasnowka 93)
3924 PKP *Ty2-1038* *+ 16.08.80*
3925 DB + 21.04.52

52 3938 ÖStB, ÖBB (Mürzzuschlag 63) + 15.07.70
3939 ÖStB, ÖBB + 20.11.54
3940 ÖStB, ÖBB + 20.11.54
3941 ÖStB, ÖBB (Knittelfeld 63) + 09.07.66 § 985.215
3942 ÖStB, ÖBB (Wien Ost 63) + 01.08.70
3943 ÖStB, ÖBB (Linz 63) + 17.02.75

---

52 3950 JDZ 33-191
3951 USSR (Be 58) (Ungeny 94)
3952 USSR (Kln 51, Li 59), (63) PKP Ty2-1305
3953 ÖStB, ÖBB (Straßhof 63) + 20.10.72
3954 USSR + 87
3955 USSR (S 53, i 57 Rosdol Schwefelkombinat) + 70
3956 USSR (Be 58), Uk
3957 CCCPc BDZ 15.217 ?
3958 PKP *Ty2-1039* *+ 15.05.90*
3959 DB + 18.10.54
52 3960 USSR, (63) CSD 555.0261 555.3261 + 73
3961 USSR (SW 48, Lw 61), (64) JZ 33-291

---

52 4045 USSR (Be 58), (62) DR (Brandenburg 70) +17.02.76
4046 BDZ 15.45
4047 BDZ 15.46 (Stanke Dimitrow 90)
4048 BDZ 15.47 (Asenowo 90)
4049 BDZ 15.48 (Asenowo 92)

---

52 4360 - 52 4373

KRENAU

12/44 - 1/45

---

52 4360 DB + 10.08.57
4361 USSR Uk 1042.287 (Rawa Russkaja 92)
4362 USSR (OK 53, Mo 56, Be 60),
(62) PKP Ty2-1220 + 85 → Vietnam
4363 USSR 1042.289 (Mamonowo 92)
4364 ÖStB, ÖBB (Selzthal 63) + 17.02.75
4365 USSR
4366 USSR (Lw 46, BL 47, Be 53, i 57 Kohle) + 69
4367 USSR (Lw 52), (63) CSSR
4368 USSR
4369 ÖStB (45) → 52 6944" → USSR
52 4370 PKP *Ty2-1040* *+ 05.03.75*
4371 USSR (Lw 50) + 01.08.73
4372 CSD 555.0106 + 69
4373 USSR Be 1042.292 (Grodno 93)

---

52 4400 - 52 4420

HENSCHEL

52 4400 - 4409 8/44
52 4410 - 4420 9/44

---

52 4400 JDZ
4401 USSR Uk 1042.293 (Wjerbka 89)
4402 JDZ 33-105
4403 JDZ 33-106
4404 USSR (Ki 50, Len 52, Okt 53, i 57 Kohlenindustrie)
4405 USSR (Be 58), Uk oder BDZ 15.229
4406 USSR 1042.295 (Kaliningrad 85) +91
4407 ÖStB + 30.06.47
4408 ÖStB, ÖBB + 27.11.57 → Ungarn
4409 USSR (Kln 51, Li 59) → PKP Ty2-1276
52 4410 USSR 1042.296 (Tapa 85) + 87
4411 USSR (Lw 46, Lw 53), (63) MAV 520,089 + 12.73
4412 ÖStB + 01.09.46
4413 USSR 1042.297 (Jelgava 85) + 88
4414 USSR (Lw 46, Li 57) i 58

52 4415 ÖStB, ÖBB (Wels 63) + 21.03.75
4416 JDZ 33-193
4417 USSR 1042.298 (Ba 85) + 89
4418 ÖStB, ÖBB (Graz 63) + 15.07.70
4419 PKP *Ty2-1042* *+ 01.08.85*
52 4420 ÖStB, ÖBB + 14.09.57 → Ungarn

---

52 4500 -

DWM / CEGIELSKI

52 4500 - 4557 9/44 - 12/44
52 4558 - 45?? 1/45 -

---

52 4500 PKP *Ty2-1175*
4501 DR 52 8094 (Lübbenau 70) + 18.11.83
4502 USSR (SW 47, MK 48, S 53, i 56 Buntmetallindustrie)
4503 DR (Engelsdorf 70) + 18.11.77
4504 USSR (Okt 53, Be 62), Pi (Szczakowa 91)
4505 DB + 04.05.52
4506 PKP *Ty2-1043* *+ 17.08.81*
4507 DR (Frankfurt/Oder 70) + 26.09.77
4508 USSR Uk
4509 DB + 20.11.58
52 4510
4511 CCCPm →CFR 150.1103 (Oravita 94)
4512 USSR (BL 47, Be 53, Lw 61),
(62) DR 52 8085 (Frankfurt /O 70) z 88 M
4513 PKP *Ty2-1044* *+ 23.06.88*
4514 USSR (Kow 46, Wnz 52, SW 53) i 56
4515 USSR (Lw 53), (63) MAV 520,081 + 04.79
4516 USSR Uk 1042.300 (Rawa Russkaja 92)
4517 PKP *Ty2-1045* *+ 28.11.79*
4518 USSR Be 1042.608 (Berestowiza 93)
4519 USSR Uk (Wjerbka 89, Rawa Russkaja 92)
52 4520 USSR (Lw 50), (64) BDZ 15.208 (Kalojanowetz 92)
4521 USSR Be 1042.302 (Wysokoje 93)
4522 USSR (La 47, Ba 53, La 56, i 59 Riga)
4523
4524 PKP *Ty2-1046* *+ 14.08.84*
4525 USSR, (63) CSD 555.0286 555.3286 + 68
4526 USSR (Lw 50), (64) BDZ 15.244 (Bodrowo 92)
4527 USSR (Lw 49, Mo 59), (64) BDZ 15.205
4528 USSR (Lw 46, Be 47, Li 62) + 01.06.80
4529 USSR Uk 1042.303 (Rawa Russkaja 92)
52 4530 DB + 15.11.57
4531 PKP *Ty2-1047* *+ 05.10.90*
4532 USSR 1042.304
4533 DR (Elsterwerda 70) + 25.08.78
4534 DR (Kamenz 70) + 22.07.77
4535 USSR (Ba 53, La 56), (63) MAV 520,5004 + 05.70
4536 PKP *Ty2-1048* *+ 13.02.79*
4537 USSR (Lw 46, Li 58, i 69 Zuckerfabrik Gorodok)
4538 PKP *Ty2-1049* *+ 15.07.82*
4539 DB + 15.11.57
52 4540 PKP *Ty2-1050* *+ 23.02.84*
4541 PKP *Ty2-1176* *+ 19.11.91*
4542 USSR 1042.305 (Krasnowka 93)
4543 USSR Be 1042.306 (Luninez 93)
4544 USSR (BL 47, Be 53) → Pi M
4545 USSR + 88
4546 USSR (Li 47, Ba 53, La 56) + 01.02.73
4547 USSR 1042.308 (Kaliningrad 85) + 89
4548 USSR (Lw 49), (63) MAV 520,090 + 01.78
4549 USSR (Lw 49), (64) BDZ 15.190
52 4550 ÖStB, ÖBB (Wiener Neustadt 63) + 27.10.78
4551 ÖStB, ÖBB (Straßhof 63) + 16.12.74
4552 ÖStB, ÖBB (Knittelfeld 63) + 10.78 M
4553 ÖStB, ÖBB (Mürzzuschlag 63) + 25.10.73
4554 USSR (OK 53, Od 56, Li 58) → PKP
4555 JDZ 33-192
4556 USSR (Od 49, OK 53, Od 56, i 56 Kohlenindustrie)
4557 ÖStB, ÖBB (St Veit/Glan 63) + 28.05.75
4558 ÖStB, ÖBB (St Veit/Glan 63) + 14.03.77
4559 ÖStB, ÖBB (Knittelfeld 63) + 24.03.77

52 4560 USSR (Li 46, La 47, Ba 53) i 72
4561 USSR Be 1042.309 (Berestowiza 93)
4562 DR (47) → 52 1452" + 19.05.76
4562" (ex 52 1452) DR (47) → USSR (Be 58)
4563 USSR (BL45, Be 47, Okt 58, Be 60), (63) Ty2-1259 + 89
4564 USSR 1042.310 (Tschernjachowsk 93)
4565 USSR Be 1042.311 (Berestowiza 93)
4566 USSR (Kow 46, 1524mm 09.47,Lw 53....
...1435mm 05.58), (64) BDZ 15.209 (Kalojanowetz 92)
4567 USSR (Radviliskis 65) 1042.312 → LG
52 4568 - 52 4573 (PKP → Ty42) ?
4574 PKP Ty42.39
52 4575 - 52 4577 (PKP → Ty42) ?
4578 USSR Uk 1042.313 (Rawa Russkaja 92)
52 4579 - 52 45?? (PKP → Ty42) ?

---

52 4731

HDZ 33-005 → DB 52 4731 + 10.08.57

---

52 4750 - 52 5124

MBA

52 4750 - 4939 1/43 - 12/43
52 4940 - 4966 1/44
52 4976 - 5124 1/44 - 11/44

---

52 4750 USSR 1042.617 (Vilnius 85)
4751 USSR (St 44 → TE-8026) # M
4752 JDZ 33-107 M
4753
4754 USSR 1042.314 (Zheleznodorozhnyj 93)
4755 NSB 4755 + 24.09.63
4756 DB + 18.10.54
4757 USSR (Kln 51, Okt 61) + 01.04.78
4758 USSR (Lw 46), (63) MAV 520,091 + 03.71
4759 USSR (i)
52 4760 USSR (Len 51, Okt 53) i 67
4761
4762 DB + 18.10.54
4763 PKP? *Ty2-405* ?
4764 ÖStB, ÖBB 152.4764 (Amstetten 63) + 01.08.70
4765 DR(o) + 46
4766 USSR (Len 51, Okt 53) i 73
4767 DB + 18.10.54
4768 USSR (Li 45, La 46, Kln 51, Okt 55, Li 62)
4769 CCCPd
52 4770 PKP
4771 USSR (SW 48, S 53, Li 58) + 01.10.84
4772 DR(w) + 12.07.48
4773 DR (Dresden 70) + 27.10.71
4774 PKP *Ty2-407* *+ 04.07.91*
4775 ÖStB, ÖBB 152.4775 (Linz 63) + 06.12.68
4776 DR + 50
4777 USSR (SW 50, La 57, i 74 Riga)
4778 JDZ 33-226
4779 DB, (51) JDZ 33-248 (Tuzla 96)
52 4780 DR (Berlin Schöneweide 70) + 30.09.71
4781 USSR (St 44, Ki 49, OK 53, Od 56, Li 58), (63) PKP
4782 PKP *Ty2-408* *+ 28.05.78*
4783 PKP *Ty2-409* *+ 31.08.84*
4784 PKP Ty2-410 (Malbork 93) + 23.07.92
4785 ÖStB, ÖBB 152.4785 (68) → GKB + 70
4786 DB + 54
4787 DR (Engelsdorf 70) + 20.04.77
4788 USSR (Od 48, OK 53, Kln 55, Be 57, Li 59, i 75 Neman)
4789 DB + 18.10.54
52 4790 PKP *Ty2-411*
4791 USSR (Klaipeda 46) + 51
4792 KrV CCCPm
4793 USSR (Od 50, OK53, Od 56, Mo 61),
(64) BDZ 15.274 (Kalojanowetz 92)
4794 DB + 18.10.54
4795 DB + 18.10.54
4796 USSR 1042.315 (Krasnowka 93)
4797 DR (Kamenz 70) + 22.02.79
4798 DR, (56) BDZ 15.92
4799 USSR (Be 58, Ba 85) 1042.316 + 87
52 4800 DB + 18.10.54
4801 USSR (Kow 50, Kln 51, Be 57) + 01.08.73
4802 USSR (Lw 46), (63) MAV 520,021 + 12.71
4803 USSR (Li 50, Ba 53, Li 56) + 01.05.72
4804 CCCPc
4805 USSR (Lw 44, Kow 46, Be 61) i 63
4806 SWDE, DB (Freiburg 50) + 18.10.54
4807
4808 DB + 18.10.54
4809 USSR (St 44) → PKP *Ty2-412* *+ 15.09.89*
52 4810 DB + 18.10.54
4811 USSR Be 1042.317 (Berestowiza 93)
4812 USSR Uk (Lw 91)
4813 USSR (BL45, Be46, Be 58) + 01.04.68
4814 JDZ 33-108
4815 USSR (Od 44, ohne Kessel) + 48
4816 PKP *Ty2-413* *+ 08.12.82*
4817 DR (Berlin Ostbahnhof 70) + 21.07.81
4818 USSR (Kln 51, Okt 61) + 01.03.70
4819 PKP *Ty2-414* *+ 09.08.90*
52 4820 USSR (BL 47, Len 52, Okt 53) + 01.06.80
4821 DR (Kamenz 70) + 17.01.72
4822 DR, (56) BDZ 15.94
4823 USSR (BL 47, Be 53, La 58) + 01.10.70
4824 CCCPd (47) → ?
4825 DR (Falkenberg 70) + 02.06.76
4826 DR (Cottbus 70) + 03.04.78
4827 DR (Hoyerswerda 70) + 20.12.79
4828 USSR (MK 53, Kln 58, Okt 61, i 71 Wjartsilja) + 73
4829 CSD 555.0119, (59) BDZ 15.117 (Kalojanowetz 92)
52 4830 NSB 4830 + 24.09.63
4831 NSB 4831 + 24.09.63
4832 NSB 4832 + 04.70
4833 NSB 4833 + 24.09.63
4834 NSB 4834 + 24.09.63
4835 NSB 4835 + 04.70
4836 NSB 4836 + 04.70
4837 NSB 4837 + 54
4838 NSB 4838 + 07.70
4839 NSB 4839 + 08.63
52 4840 PKP *Ty2-415* *+ 08.07.86*
4841 CSD 555.0121, 555.3121 + 68
4842 CSD 555.077, 555.3077 + 73
4843 CSD, BDZ 15.122 (Bodrowo 92)
4844 USSR (Ba 53, La 56, Kln 58, Okt 61) + 01.05.74
4845 USSR (La 47, Ba 53, La 56) + 01.09.70
4846 USSR (Lw 44, Od 48, Lw 49) + 01.05.71
4847 USSR (Be 47, MK 48, Li 57) + 01.01.74
4848 PKP *Ty2-416* *+ 15.12.87*
4849 USSR 1042.319 (Bagrationowsk 93)
52 4850 PKP *Ty2-417*
4851 JDZ 33-109
4852 PKP *Ty2-418* *+ 01.12.86*
4853 USSR Li 1042.320 (Radviliskis 92) + 92
4854 PKP *Ty2-419* *+ 05.10.90*
4855 TCDD 56516 (Balikesir 81, Halkapinar 95)
4856 TCDD 56517 (Afyon 81, Halkapinar 95)
4857 TCDD 56518 (Afyon 81, Afyon 95)
4858 TCDD 56519 (Bilecik 69)
4859 TCDD 56520 (Usak 81, Usak 95)
52 4860 TCDD 56521 (Izmir 81, Alasehir 95)
4861 TCDD 56522 (Tavsanli 81, Afyon 95)
4862 TCDD 56523 (Balikesir 81) M
4863 TCDD 56524 (Usak 81, Usak 95)
4864 TCDD 56525 (Izmir 75)

52 4865 DB + 18.10.54
4866 PKP *Ty2-420* + *28.12.89*
4867 ÖStB, ÖBB 152.4867, (70) → GKB + 80 M
4868 DB + 18.10.54
4869 DR (Berlin Pankow 69) + 17.03.70
52 4870 DR, (56) BDZ 15.99 (Bodrowo 92)
4871 PKP *Ty2-421* + *23.07.92*
4872 DR(w) + 46
4873 PKP *Ty2-422* + *05.10.90*
4874 DB + 18.10.54
4875 DR (Elsterwerda 70) + 16.09.77
4876 DR 52 8059 (Altenburg 70) + 19.08.88
4877 PKP *Ty2-423* + *04.11.89*
4878 USSR (Be 47, SW 48, S 53, i 57 Kohlenindustrie)
4879 USSR (Od 56, Li 58), (62) DR (Kamenz 70) + 01.02.77
52 4880 JDZ 33-194
4881 PKP *Ty2-424* + *10.11.87*
4882 USSR 1042.321 (Mamonowo 93)
4883 PKP *Ty2-425* + *14.02.80*
4884 USSR (Ba 53, Li 56), (63) PKP
4885 USSR, (62) CSD 555.0228 555.3228 + 67
4886 PKP *Ty2-426*
4887 DB, (51) JDZ 33-249
4888
4889 DB + 18.10.54
52 4890
4891 USSR (BL 50, Be 53),
(63) PKP Ty2-1340 + 05.10.90 → Dsp, (Szczecinek 96)
4892 DR (Wittenberg 70) + 24.10.75
4893 DR (Falkenberg 70) + 20.05.81
4894 PKP *Ty2-427* + *01.02.75*
4895 CSD
4896 DR 52 8154 (Hoyerswerda 70) M
4897 USSR (Be 58, Ba 85) 1042.322 + 88
4898 PKP *Ty2-428* + *21.07.83*
4899 DB + 18.10.54
52 4900 DR 52 9900 (Senftenberg 70) M
4901 DR 52 8195 (Templin 70) M
4902 DR (Wustermark 70) + 22.02.79
4903 USSR (Jelgava 84) + 85
4904 USSR (i) Workuta 52
4905 PKP *Ty2-429* + *15.05.86*
4906 PKP
4907 DR 52 8142 (Hoyerswerda 70) z 07.87 M
4908 DR (Stralsund 70) + 01.02.77
4909 PKP *Ty2-431* + *25.03.87*
52 4910 CCCPd
4911 DR (Halle 70) + 31.01.79
4912 DR (Dresden 70) + 23.04.75
4913 USSR (SW48, Od 57, Be62), (63) CSD 555.0255 + 70
4914 USSR (i) Lesnaja 65
4915 DB (VA Göttingen 50) + 18.10.54
4916 USSR 1042.324 (Wjerbka 89) + 88
4917 PKP
4918 USSR (Es 57, Lw 60), (63) MAV 520,017 + 06.78
4919 USSR (OK 53, Od 56, Li 58), → PKP Ty2-1281 + 90
52 4920 USSR (SW 47, Od 57, Mo 61),
(64) BDZ 15.269 (Karnobat 92)
4921 DR (Stendal 70) + 02.06.76
4922 DB + 18.10.54
4923 PKP *Ty2-433*
4924 DR (Berlin Schöneweide 70) + 04.90 M
4925 USSR (Be 50), → Pi (Szczakowa 91)
4926 DR (Frankfurt/Oder 70) v 01.01.85 x 09.87
4927 USSR (Lw 49, Kln 51, Okt 61) + 01.06.72
4928 PKP *Ty2-434* + *18.02.87*
4929 NSB 4929 + 04.70

52 4930 PKP *Ty2-435*
4931 DB + 18.10.54
4932 PKP *Ty2-436* + *15.01.77*
4933 PKP *Ty2-437* + *29.11.90*
4934 CCCPd
4935 DR(o) + 47
4936 JDZ 33-110 M
4937 USSR (Lw 53), (64) BDZ 15.202 (Bodrowo 92)
4938 USSR (Ba 53, Li 56, Mo 60), Export
4939 DB + 18.10.54
52 4940 JDZ 33-111
4941 USSR (MK 53, Li 58), (63) PKP Ty2-1271
4942 USSR (Lw 51), (64) BDZ 15.183 (Asenowo 92)
4943 ÖStB, ÖBB + 10.08.59 § 985.212
4944 DR 52 8012 (Cottbus 70) M
4945 DB + 18.10.54
4946 USSR 1042.325 (Vilnius 85)
4947 PKP *Ty2-438* + *23.01.76*
4948 USSR, (62) CSD 555.0247 555.3247 + 68
4949 USSR (SW 50, Len51, Okt 53, i 57 Waldindustrie)
52 4950 USSR (Lw 53), (64) BDZ 15.192
4951 USSR (Li 45) i 51
4952 DR (Frankfurt/Oder 70) + 19.05.76
4953 DR + 25.05.54
4954 DR (Engelsdorf 69) + 24.03.77
4955 DR (Cottbus 70) + 03.11.69
4956 USSR (Kln 51, Be 58), (63) PKP Ty2-1349+ 15.05.90
4957 USSR (Kritschew 47) i 53
4958 DB + 18.10.54
4959 DB + 13.12.50
52 4960 DB + 18.10.54
4961 PKP *Ty2-1051* + *11.04.84*
4962 USSR 1042.326 (Ba 85) + 90
4963 PKP *Ty2-439* + *17.03.90*
4964 PKP *Ty2-440* + *28.12.89*
4965 DR (Engelsdorf 70) + 02.12.75
4966 DR (Karl-Marx-Stadt 70) z 87 M

4976 BDZ 15.49
4977 BDZ 15.50 (Bodrowo 92)
4978 BDZ 15.51 (Bodrowo 92)
4979 JDZ 33-220
52 4980 BDZ 15.52 (Bodrowo 92)
4981 BDZ 15.53 (Bodrowo 92)
4982 JDZ 33-112
4983 BDZ 15.54
4984 JDZ 33-227 M
4985 JDZ 33-055
4986
4987 USSR (Okt 53, Mo 61), (63) MAV 520,011 + 04.64
4988 CCCPc
4989 USSR (La 47, Ba 53, La 56) i 68
52 4990 PKP *Ty2-1052* + *18.07.78*
4991 PKP Ty2-160 (Nowe Skalmierzyce 91) + 15.05.90
4992 PKP *Ty2-1053* + *30.11.86*
4993 PKP *Ty2-1054* + *18.07.72*
4994 DR (Templin 70) + 16.09.77
4995 PKP Ty2-1055 + 01.87 M
4996 DR 52 8177 (Berlin Schöneweide 70) z 91 M
4997 DB + 18.10.54
4998 DR (Dresden 70) v 10.11.78
4999 DB + 18.10.54

52 5000 USSR (OK 53, Od 56, Be 58), (63) PKP
5001 PKP *Ty2-1056* *+ 16.04.86*
5002 PKP
5003 PKP *Ty2-1058* *+ 31.01.88*
5004 DR (Kamenz 70) v 27.07.77
5005 PKP *Ty2-1059* *+ 05.10.90*
5006 USSR Be 1042.327 (Wysokoje 93)
5007 DR (Leipzig Süd 70) + 15.03.73
5008 DR (Falkenberg 70) + 30.07.75
5009 PKP *Ty2-442* *+ 27.11.86*
52 5010 USSR (Lesnaja 64)
5011 PKP *Ty2-1060*
5012 DR 52 8030 (Lübbenau 70) Dsp 88 M
5013 DB + 18.10.54
5014 DR (Falkenberg 70) + 25.08.78
5015 DR 52 8080 (Lübbenau 70) z 88 M
5016 PKP *Ty2-1061* *+ 26.03.91*
5017 USSR (Li 47) i 51
5018 DR 52 8029 (Frankfurt/Oder 70) z 91 M
5019 USSR (Li 47, Ba 53, Li 56) + 01.07.78
52 5020 DB + 18.10.54
5021 USSR (BL 45, Be 46, SW 48, Lw 61), (66) JZ 33-324
5022 USSR (Be 53, Mo 60), (63) MAV 520,031 + 12.79
5023 PKP *Ty2-1062* *+ 04.06.76*
5024 DR (Berlin Schöneweide 70) + 03.04.78
5025 DB + 18.10.54
5026 USSR, (63) CSD 555.0260 555.3260 + 72
5027 USSR Be 1042.328 (Wysokoje 93)
5028 USSR (Kln 53, Be 57, Mo60), BDZ 15.148
5029 JDZ 33-113
52 5030 DR (Frankfurt/Oder 70) + 22.02.79
5031 DB + 18.10.54
5032 NSB 5032 + 04.70
5033 USSR Be (Luninez 93)
5034 JDZ 33-114
5035 USSR (Li 47, Kln 53, Okt 55) + 01.07.74
5036
5037 USSR (Be 53, La 58, Lw 59) i 67
5038 USSR, (63) CSD 555.0273 555.3273 + 68
5039 USSR (Okt 53, La 57, Kln 58, Lw 60), (64) JZ 33-275
52 5040
5041 SWDE, DB (Mainz 50) + 18.10.54
5042 DB + 18.10.54
5043 DB + 18.10.54
5044 USSR (Be 50, La 58, Li 59), (63) PKP Ty2-1392
5045 (CSD 555.0138)?
5046 DR (Halle 70) + 25.08.78
5047 DB + 18.10.54
5048 USSR (Be 53), (63) PKP Ty2-1262 + 23.02.90
5049 USSR (Ba 53, Li 56, OK 64) + 01.03.79
52 5050 PKP *Ty2-1063*
5051 SWDE, DB (Freudenstadt 50) + 18.10.54
5052 ÖStB, ÖBB (Knittelfeld 63) + 26.03.73
5053 DB + 18.10.54
5054 USSR (Lw 49) i 51
5055 USSR (SW48, Lw61), (62) DR (Jüterbog 70)+ 23.01.79
5056 JDZ 33-115
5057 ÖStB, ÖBB + 23.12.57 → Ungarn
5058 PKP *Ty2-1064* *+ 15.05.90*
[Nach Griebl / Wenzel: 13.02.45 52 5058 → 52 5098]
5059 SAAR, (57) DB + 20.11.58
52 5060 PKP *Ty2-1065* *+ 09.12.85*
5061 DB (Herford 50) + 18.10.54
5062 PKP *Ty2-1066* *+ 15.05.90*
5063 PKP *Ty2-1067* *+ 18.11.86*
5064 CCCPd
5065 PKP *Ty2-1068* *+ 17.04.85*
5066 USSR (Okt 53, Lw 61), (64) JZ 33-276
5067 DR (Falkenberg 70) + 30.07.75
5068 USSR 1042.330 (Valga 89) + 90
5069 USSR (Lw 59), (63) MAV 520,043 + 05.79
52 5070 USSR (Be 53), (63) PKP Ty2-1336 + 04.11.89
5071 PKP Ty2-370 + 26.07.89
5072 USSR (Lw 48), (66) JZ 33-340
5073 USSR Be 1042.609 (Luninez 93)
5074 USSR, BDZ 15.161
52 5075 SWDE,DB (Betzdorf 50) + 18.10.54
5076 PKP *Ty2-1070* *+ 20.09.89*
5077 USSR (Be 50, Kln 54, i 56 Kohlenindustrie)
5078 USSR 1042.331 (Kaliningrad 92)
5079 USSR (Rezekne 58) 1042.332 + 86
52 5080 DB (Löhne 50) + 18.10.54
5081 NSB 5081 + 24.09.63
5082 USSR (Li 47, Ba 53, Es 56) + 01.02.73
5083 USSR (Be 53, Okt 58, Be 60), (62) PKP Ty2-1219
5084 USSR (Lw 53), (64) BDZ 15.178 ?
5085 USSR (Kow 49, Lw 53, Li 57) i 58
5086 PKP *Ty2-1071* *+ 20.02.73*
5087 PKP *Ty2-1072* *+ 14.06.89*
5088 DR (Falkenberg 70) + 22.02.79
5089 PKP *Ty2-1073* *+ 23.02.90*
52 5090 USSR (BL 45-47, Lw 48) i 65
5091 DR (Engelsdorf 70) + 19.08.75
5092
5093 USSR (Ba 53, La 56, Be 59), (63) PKP Ty2-1246
5094 PKP *Ty2-1074* *+ 19.11.91*
5095 PKP *Ty2-1075*
5096 USSR (Be 58) + 86 → Vietnam x 94 ?
5097 PKP *Ty2-1076* *+ 20.04.90*
5098 DR [Siehe 52 5058] (Zittau 70) + 29.07.76
5099 PKP *Ty2-1077* *+ 01.06.82*
52 5100 PKP *Ty2-1078*
5101 PKP *Ty2-1079* *+ 09.10.79*
5102 PKP *Ty2-1080* *+ 04.11.89*
5103
5104 PKP *Ty2-1081* *+ 15.10.65*
5105 USSR (62) CSD 555.0237 555.3237 + 68
5106 PKP *Ty2-1082* *+ 05.07.91*
5107 PKP *Ty2-1083* *+ 04.11.89*
5108 USSR (Okt 53, Lw 61), (64) JZ 33-270
5109 JDZ 33-195
52 5110 USSR (Li 47, Ba 53, Li 56) i 64
5111 DR (Bautzen 70) + 14.05.79
5112 USSR (OK 53, Od 56, Mo58), BDZ 15.175
5113 NSB 5113 + 11.08.65
5114 PKP *Ty2-1084* *+ 29.04.91*
5115 USSR (BL45, SW 48, S 53, i 56 Kohlenindustrie)
5116 NSB 5116 + 20.08.64
5117 PKP *Ty2-443* *+ 31.01.81*
5118 DR 52 8061 (Frankfurt/Oder 70) + 11.04.74
5119 DB + 18.10.54
52 5120 PKP *Ty2-1085*
5121 DR 52 8052 (Lübbenau 70) + 09.08.88
5122 DB + 18.10.54
5123 PKP Ty2-1086 (Wolsztyn 95)
5124 USSR (BL46, Li 47, Okt 53, i 57 Kohlenindustrie)

---

52 5125 - 52 5374

KRENAU

52 5125 - 5286 4/43 - 12/43
52 5287 - 5374 1/44 - 6/44

---

52 5125 PKP *Ty2-444* *+ 15.09.89*
5126 DR (Bautzen 70) + 20.08.76
5127 USSR (Okt 53, Mo 61), (63) MAV 520,052 + 05.77
5128 USSR (Lw 59), (63) MAV 520,082 + 05.79
5129 USSR + 88
52 5130 USSR (NK45, La50, Ba 53, La56, Lw61), (62) DR + 68
5131 USSR (St 44) → BDZ 15.210
5132 PKP *Ty2-445* *+ 25.08.88*
5133 NSB 5133 + 11.01.51
5134 PKP *Ty2-446* *+ 16.08.93*
5135 USSR (Li 48, Ba 53, Es 56) + 01.04.73
5136 PKP *Ty2-447* *+ 02.06.87*
5137 DR (Zittau 70) + 19.06.86
5138 PKP *Ty2-448* *+ 20.06.88*
5139 PKP *Ty2-449* *+ 10.11.87*

52 5140	DR (Wittenberg 70)	+ 24.10.78	
5141	USSR (Be 53, La 58, Li 59), (63) PKP Ty2-1397	+ 88	
5142	DR (Frankfurt/Oder 70)	+ 19.05.76	
5143	CSD 555.0110	+ 71	
5144	USSR (OK 53, Ba 55, Li 56)	+ 01.09.75	
5145	USSR (62) CSD 555.0218 555.3218	+ 69	
5146	USSR (Kln 51, Li 58)	i 75	
5147	USSR (Kln 53, Be 57, Li 58), (63) PKP Ty2-1268	+ 90	
5148	USSR (Kow 47, Wnz 52, SW 53, i 57 Kohlenindustrie)		
5149	DB	+ 10.08.57	
52 5150	USSR Be 1042.335 (Luninez 93)		
5151	USSR (SW53, La 56, Mo 61), (64) BDZ 15.252		
5152	ÖStB, ÖBB + 31.07.57 → Ungarn		
5153	CCCPc		
5154	DR (Jüterbog 70)	+ 17.01.72	
5155	DB	+ 07.02.52	
5156	USSR, (63) MAV 520,053 520,034 #	+ 84	M
5157	USSR (Be 58, Ba 85) 1042.336	+ 85	
5158	DR(w)	+ 47	
5159	USSR (La 48, Ba 53, La 56)	+ 01.10.71	
52 5160	PKP Ty2-706 (Nysa 87)	+ 23.02.90	
5161	PKP *Ty2-451*	*+ 18.11.86*	
5162	PKP *Ty2-1185*	*+ 25.10.85*	
5163	DB	+ 10.08.57	
5164			
5165	USSR (Radviliskis 55) 1042.337	+ 89	
5166	USSR (La 47, Kln 51, La 57)	+ 01.04.69	
5167	PKP *Ty2-452*	*+ 16.06.89*	
5168	PKP *Ty2-453*	*+ 10.09.88*	
5169	DB	+ 07.02.52	
52 5170	USSR (Lw 53, Ba 54, La 56 Mo 61),	(64) BDZ	
5171	PKP *Ty2-454*		
5172	DB	+ 20.11.58	
5173	DB	+ 20.07.59	
5174	USSR (Be 53, Lw 62), (63) CSSR		
5175	DR(o) KrV	+ 47	
5176	USSR (Ki 48, OK 53, Mo56), BDZ 15.149	(Asenowo 92)	
5177	USSR (La 47, Ba 53, Es 56)	+ 01.05.78	
5178	DB	+ 10.08.57	
5179	JDZ 33-116		
52 5180	DB	+ 10.08.57	
5181	USSR (62) CSD 555.0214 555.3214	+ 71	
5182	USSR (SW 50, La 57, i 58 Riga)		
5183	USSR (Be 53, Lw 62), (63) CSSR		
5184	ÖStB, ÖBB	+ 28.10.54	
5185	DB	+ 10.08.57	
5186	USSR (Lw 53, Be 58), (62) DR (Zittau 70)	+ 12.07.76	
5187	ÖStB, ÖBB (Knittelfeld 63)	+ 20.03.73	
5188	USSR (Lw 46)		
5189	PKP *Ty2-455*	*+ 29.04.91*	
52 5190	DR 52 8082 (Lübbenau 70)	z 88	M
5191	DR	v 27.07.76	
5192	USSR (Ba 53, La 56, Lw 59), (63) CSSR		
5193	USSR (OK 53, Od 56, Li 58, i 58 Kohle. Makeewka)		
5194	CSD 555.0111		
5195	PKP *Ty2-456*	*+ 26.08.88*	
5196	PKP *Ty2-457*	*+ 09.02.88*	
5197	USSR (Kln 51, Okt 55, Li 62)	+ 01.05.74	
5198	PKP Ty2-458 (Jaworzyna 87)	+ 17.03.90	
5199	DR (Salzwedel 70)	+ 04.01.77	
52 5200	USSR 1042.338 (Kowel 86)		M
5201	USSR (Li 48, Okt 53, Es 56)	+ 01.08.76	
5202	PKP *Ty2-459*	*+ 08.07.86*	
5203	USSR (Wnz 52, SW 53, La 56, i 56 Kohlenindustrie)		
5204	DB	+ 10.08.57	
5205	PKP *Ty2-460*	*+ 92*	
5206	USSR (OK 53, Mo56, Lw 59), (63) CSSR		
5207	DR 52 8014 (Görlitz 70)	z 07.87	M
5208	USSR (Lw 52), (64) BDZ		
5209	DR(o) → 52 5022"	+ 12.47	
52 5210	USSR (Lesnaja 58)		
5211	ÖStB, ÖBB (Wien Ost 63)	+ 29.11.77	
5212	DR (Bautzen 70)	+ 06.09.75	
5213	DR	+ 02.01.69	
5214	PKP *Ty2-461*	*+ 28.11.89*	
52 5215	DR (Frankfurt/Oder 70)	+ 12.02.80	
5216	USSR, (63) CSD 555.0280 555.3280	+ 66	
5217	USSR Uk 1042.339 (Chyrow 94)		
5218	DB	+ 10.08.57	
5219	USSR (Od 56, Mo 60), (63) MAV 520,004	+ 12.74	
52 5220	PKP		
5221	USSR (Gusew 92) 1042.340	+ 91	
5222	USSR (SW50, Li 58), (62) DR (Dresden 70)	+ 16.01.75	
5223	USSR (Li 50, Ba 53, Li 56)	+ 01.08.74	
5224	JDZ 33-117		
5225	DR(o)	+ 04.46	
5226	USSR (Ki 48, Wnz 51, SW 53)	i 56	
5227	SAAR, (57) DB	+ 15.11.57	
5228	DR, (59) BDZ 15.135 ?		
5229	USSR (Kow 45, ... , Li 48, Ba53)	+ 01.11.77	
52 5230	USSR, (63) CSD 555.0289 555.3289	+ 71	
5231	SWDE,DB (Freudenstadt 50)	+ 15.11.57	
5232	DR (Engelsdorf 70)	+ 30.07.75	
5233	DR (Frankfurt/Oder 70)	+ 07.08.79	
5234	DR	+ 21.02.69	
5235	PKP *Ty2-463*		
5236	DB	+ 20.07.59	
5237	USSR (Ba 53, La 56, Li 58), → PKP		
5238	DR	v 13.10.69	
5239	DR (Stralsund 70)	+ 20.07.78	
52 5240	PKP *Ty2-464 + 28.03.85 (85) → Vietnam*		
5241	DR (Pasewalk 70)	+ 01.03.71	
5242	PKP *Ty2-465*	*+ 16.06.75*	
5243	DR 52 8041 (Altenburg 70)		M
5244	DR(w) (10/45) → DR(o) → 52 7559"		
5245	PKP *Ty2-466*	*+ 20.09.89*	
5246	PKP *Ty2-467*	*+ 08.08.75*	
5247	PKP *Ty2-468*		
5248	USSR Be 1042.341 (Luninez 93)		
5249	PKP *Ty2-469*	*+ 15.05.90*	
52 5250	DR 52 8165 (Stendal 70)	z 87	M
5251	DR	+ 20.12.51	
5252	DR(o)	+ 47	
5253	DB	+ 10.08.57	
5254	PKP *Ty2-470*	*+ 10.02.87*	
5255	USSR (Kln 51, Okt 55, Mo 61), (64) BDZ		
5256	USSR (Be 58)		
5257	USSR (Od 50, OK 53, i 56 Kohlenindustrie)		
5258	DB	+ 20.07.59	
5259	USSR 1042.343 (Riga 85)	+ 90	
52 5260	DR 52 8101 (Berlin Schöneweide 70) #		
5261	DB	+ 20.07.59	
5262	PKP *Ty2-1189*	*+ 06.07.81*	
5263	DB	+ 20.07.59	
5264	PKP *Ty2-471*	*+ 08.07.86*	
5265	USSR Be 1042.344 (Osipowitschi 93)		
5266	DB	+ 07.02.52	
5267	PKP *Ty2-472*	*+ 26.03.91*	
5268	USSR (Lw 66)		
5269	DB	+ 20.07.59	
52 5270	FS, (48) DB	+ 10.08.57	
5271	USSR Uk (Wjerbka 89)		
5272			
5273	DB	+ 10.08.57	
5274	DR 52 8038 (Cottbus 70)		M
5275	DB (Hanau 50)	+ 07.02.52	
5276	USSR (Okt 53, Mo 60), BDZ 15.168		
5277	USSR Be 1042.346 (Wysokoje 93)		
5278	DR 52 9278 (Senftenberg 70)	+ 11.10.76	
5279	PKP *Ty2-473*	*+ 16.12.91*	
52 5280	DB	+ 07.02.52	
5281	CSD 555.0107		
5282	USSR (Kln 51, Okt 55)		
5283	USSR (Ki 50, Len 53, Okt 53, Be 58)		
	(62) DR (Brandenburg 70)	+ 04.01.77	
5284	DB	+ 10.08.57	
5285	PKP *Ty2-474*	*+ 04.05.81*	
5286	USSR, (63) CSD 555.0295 555.3295	+ 72	
5287	DR (Wittenberg 70)	+ 06.09.84	
5288	PKP *Ty2-475*	*+ 10.04.92*	
5289	PKP *Ty2-476*	*+ 27.08.88*	

52 5290	USSR 1042.347 (Kaliningrad 85)	+ 89
5291	USSR (SW50, Be 55), (62) PKP Ty2-1233	+ 29.11.90
5292	DB	+ 20.07.59
5293	USSR 1042.348 (Radviliskis 85)	+ 87
5294	USSR (SW 51)	i 56
5295	PKP *Ty2-477*	
5296	PKP *Ty2-478*	+ *20.02.74*
5297	DR(o)	+ 47
5298	PKP *Ty2-479*	
5299	USSR (Li 50, Ba 53, La 56)	+ 01.03.77
52 5300	USSR (SW 50, i 56 Kohlenindustrie)	
5301	USSR (Kln 53, Be 57, Mo 59), (64) BDZ 15.230	
5302	USSR (Be 50), (62) PKP Ty2-1228	
5303	USSR (Li 49, Ba 53, La 56)	+ 01.02.73
5304	PKP *Ty2-1087*	+ *28.08.78*
5305	DB	+ 10.08.57
5306	DR 52 8189 (Eberswalde 70)	z 12.87 M
5307	USSR (Lw 59), (63) MAV 520,032	+ 11.77
5308	SWDE,DB (Freudenstadt 50)	+ 25.04.58
5309	USSR (Be 53) → Pi (Szczakowa 91)	
52 5310	ÖStB, ÖBB + 31.07.57 → Ungarn	
5311	PKP *Ty2-1088*	
5312	USSR	+ 87
5313	USSR (La 47, Kln 51, Okt 55)	i 73
5314	USSR Be 1042.350 (Wysokoje 93)	
5315	DR 52 8141 (Brandenburg 70)	M
5316	PKP *Ty2-1089*	+ *07.06.91*
5317	PKP *Ty2-1090*	
5318	USSR (BL 50, Be 53, i 57 Kohle)	+ 61
5319	USSR 1042.351 (Mamonowo 93)	
52 5320	PKP Ty2-1146 (Andrzejowka 91)	+ 16.09.91
5321	NSB 5321	+ 11.01.51
5322	PKP *Ty2-1092*	
5323	DB	+ 10.08.57
5324	PKP *Ty2-1093*	+ *26.04.76*
5325	PKP *Ty2-1094*	+ *08.12.79*
5326	USSR (Kln 53, Be 57, Li 58), (63) PKP	Ty2-1365 ?
5327	CSD 555.0100	
5328	PKP *Ty2-1095*	+ *21.09.84*
5329	PKP *Ty2-1096*	
52 5330	USSR (Li 45, La 47, Ba 53, La 56)	+ 01.06.77
5331	PKP *Ty2-481*	+ *15.01.77*
5332	DR (Leipzig Süd 70)	+ 20.02.75
5333	PKP *Ty2-1097*	
5334	SWDE, DB (Worms 50)	+ 25.04.58
5335	DB	+ 10.08.57
5336	PKP Ty2-667 (Chojnice 90)	+ 07.02.91
5337	CCCPd (47) → ?	
5338	USSR 1042.352 (Tapa 85)	+ 88
5339	PKP *Ty2-1099*	+ *06.07.83*
52 5340	PKP *Ty2-1100*	+ *04.06.84*
5341	DB (Osnabrück Vbf 50)	+ 15.11.57
5342	USSR (Li 45, Kln 51, Mo 59), (64) BDZ 15.256	
5343	USSR Be 1042.353 (Wysokoje 93)	
5344	USSR, (62) CSD 555.0227 555.3227	+ 68
5345	DR	+ 51
5346	PKP Ty2-1101 (Chabowka 91)	+ 07.06.91
5347	DB	+ 21.04.52
5348	USSR (OK53, Mo56, Be 60), (62) PKP Ty2-1230	
5349	USSR (Okt 53, Lw 61), (64) JZ 33-292	
52 5350	PKP *Ty2-1102*	+ *25.03.87*
5351	PKP *Ty2-1103*	+ *22.08.77*
5352	DB	+ 20.11.58
5353	CSD ?	
5354	DR (Görlitz 70)	+ 06.06.77
5355	DR (Halle 70)	+ 31.01.79
5356	PKP Ty2-1104 (Tarnowskie Gory 91)	
5357	ÖStB, ÖBB + 17.05.57 → Ungarn	
5358	USSR Be 1042.354 (Wysokoje 93)	
5359	SAAR, (57) DB	+ 15.11.57
52 5360	USSR Uk	
5361	DB	+ 10.08.57
5362	USSR (Lw 46, Kln 51, Be 58), (63) PKP	
5363	USSR (Be 50, Mo 61), (64) BDZ 15.258	
5364	DB ?	
52 5365	SWDE, DB (Mainz 50)	+ 25.04.58
5366	DB	+ 15.11.57
5367	PKP *Ty2-1105*	+ *08.12.82*
5368	DB	+ 21.04.52
5369	USSR (Radviliskis 55) 1042.356	+ 88
52 5370	PKP *Ty2-1106*	+ *23.02.90*
5371	NSB 5371	+ 04.70
5372	DR(o) (45) → 52 063" → USSR, (64) JZ 33-296 52 5372" (ex 52 063) DR (Salzwedel 70)	+ 20.07.78
5373	USSR (Ki 49, Wnz 51, SW 53, i 56 Kohlenindustrie)	
5374	DR 52 8196 (Templin 70)	z 07.87 M

---

52 5375 - 52 5874

SCHICHAU

52 5375 - 5769 1/43 - 12/43
52 5770 - 5874 1/44 - 4/44

---

52 5375	DR, (56) BDZ 15.93 (Asenowo 90)	
5376	DB	+ 18.10.54
5377	DR (Leipzig Süd 70)	+ 22.07.77
5378	DB	+ 18.10.54
5379	USSR (OK 53, Ba 54, Es 56)	+ 01.10.74
52 5380	USSR (Dno 83)	+ 86
5381	PKP *Ty2-482*	+ *05.10.90*
5382	DR	+ 05.06.68
5383	USSR (Klaipeda 46)	+ 51
5384	USSR (Od 44, i 45 Buntmetallindustrie)	
5385	USSR (OK 53, Mo 56), (64) BDZ	
5386	PKP *Ty2-483*	+ *01.08.85*
5387	PKP *Ty2-484*	+ *18.07.72*
5388	USSR 1042.358 (Kaliningrad 85)	+ 89
5389	USSR (Li 45, La 47, Ba 53, La 56)	+ 01.06.71
52 5390	USSR (Li 45, SW 48, Li 56)	+ 01.06.70
5391	PKP *Ty2-485*	
5392	USSR (SW48, La 56), (63) MAV 520,5005	+ 08.71
5393	DB	+ 18.10.54
5394	USSR (SW 50, S 53, i 56 Kohlenindustrie)	
5395	PKP *Ty2-486*	+ *20.12.87*
5396	DB	+ 18.10.54
5397	NSB 5397	+ 20.08.64
5398	PKP *Ty2-487*	+ *04.11.89*
5399	USSR (BL 45)	
52 5400	DB	+ 54
5401	USSR (St 44)	
5402	DB	+ 54
5403	DR	+ 21.12.53
5404	DB	+ 54
5405	USSR (La 50, Ba 53, La 56)	+ 01.02.75
5406	USSR (OK 53, Ba 55, La 56)	+ 01.07.72
5407	USSR (S 44)	
5408	USSR Be 1042.359 (Wysokoje 93)	
5409	JDZ 33-118	
52 5410	DR (Elsterwerda 70)	v 20.06.77
5411	PKP *Ty2-488*	+ *20.12.82*
5412	USSR (Kln51, Okt 61, i 71 Metallwerk Wjartsilja Karelia)	
5413	DR (Wustermark 70)	+ 25.10.74
5414	DR, (56) BDZ 15.103	
5415	USSR (Kaunas 55) 1042.360	M
5416	USSR (Ba 53, La 56, Lw 62), (66) JZ 33-337	
5417	CCCPd	
5418	CCCPc	
5419	USSR (Li 46, Ba 53, La 56)	+ 01.10.72
52 5420	USSR 1042.361 (Radviliskis 85)	+ 90
5421	USSR (OK 53, Mo 56, Be 60), (63) PKP	
5422	USSR (Kln 52, Okt 55, Lw 62), (66) JZ 33-329	M
5423	SWDE, DB (Freiburg 50)	+ 18.10.54
5424	USSR Uk 1042.362	
5425	USSR 1042.363 (Mamonowo 93)	
5426	USSR (Kow 46, Lw 53, La 57)	+ 61
5427	USSR (BL47, Len 52, Okt 53)	+ 01.03.77
5428	PKP *Ty2-489*	
5429		

52 5430	CSD 555.094 555.3094	+ 69
5431	DB	+ 18.10.54
5432	PKP *Ty2-490*	*+ 16.01.79*
5433	DB	+ 18.10.54
5434	DR, (56) BDZ 15.87	
5435		
5436	PKP *Ty2-491*	*+ 15.09.89*
5437	USSR (Kow 50, Kln 51, Okt 61)	+ 01.04.73
5438	PKP *Ty2-492*	
5439		
52 5440	PKP *Ty2-493*	*+ 23.02.90*
5441	USSR (SW 50, i 60 Wagenwerk Darniza)	+ 66
5442	SWDE, DB (Betzdorf 50)	+ 18.10.54
5443	USSR (Od 44) Ohne Kessel	+ 48
5444	USSR (Od 48, OK 53, Od 56, Lw 62), (64) JZ	
5445	DB	+ 18.10.54
5446	USSR (Be 58)	+ 89
5447	DR (Salzwedel 70)	+ 19.05.76
5448	DR (Engelsdorf 70)	M
5449	DR(o) KrV	+ 04.46
52 5450	CCCPd	
5451	USSR (Be 50)	+ 01.06.72
5452	USSR (La 47, Ba 53, La 56)	+ 01.12.74
5453	USSR (Kow 50, Wnz 52, SW 53, i 56 Kohlenindustrie)	
5454	USSR	
5455	PKP *Ty2-494*	*+ 31.01.88*
5456	PKP *Ty2-495*	*+ 03.11.89*
5457	DR (Frankfurt/Oder 70)	+ 16.05.77
5458	USSR (Vilnius 56) 1042.366	+ 89
5459	USSR Be 1042.367 (Wysokoje 93)	
52 5460	PKP *Ty2-496*	*+ 21.01.87*
5461	USSR (Kow 50, Kln 51, Okt 61)	i 64
5462	USSR (Kln 53, Be 57, SW 59)	i 61
5463	USSR (MK53, Be 55, Lw 62), (64) JZ 33-304	
5464	USSR, (63) CSD 555.0276 555.3276	+ 71
5465	USSR 1042.368 (Kaliningrad 85)	+ 87
5466	USSR (W 51, MK 53, Be 55)	i 64
5467	PKP *Ty2-497*	*+ 15.01.86*
5468	DR (Kamenz 70)	+ 24.03.77
5469	PKP *Ty2-498*	*+ 09.02.78*
52 5470	USSR + 86 → Vietnam	x 94 ?
5471	PKP *Ty2-499*	*+ 28.11.79*
5472	USSR (OK53, Od 56, Be 62), → Pi (Szczakowa 91)	
5473	USSR (Li 47, Ba 53, Li 56)	+ 01.05.71
5474	DB	+ 18.10.54
5475	DR (Wittenberg 70)	+ 20.08.76
5476	PKP *Ty2-500*	*+ 08.06.72*
5477	USSR	
5478	USSR	
5479	DB	+ 01.06.53
52 5480	PKP *Ty2-501*	*+ 14.07.86*
5481	DR (Jüterbog 70)	+ 21.02.77
5482	PKP *Ty2-502*	*+ 08.02.85*
5483	USSR 1042.372 (Zheleznodorozhnyj 93)	
5484	PKP *Ty2-503*	*+ 17.03.90*
5485	USSR (Li 45)	
5486		
5487	DR(w)→DR(o) →DR → (56) BDZ 15.88	
5488	DR (Falkenberg 70)	v 11.78
5489	DB	+ 18.10.54
52 5490	USSR + 86 → Vietnam	x 94 ?
5491	USSR (SW48, Od 57, Mo 61), (64) BDZ 15.260	
5492	PKP *Ty2-504*	*+ 13.09.88*
5493	PKP *Ty2-505*	*+ 26.07.89*
5494	PKP *Ty2-506*	*+ 89*
5495	DB	+ 18.10.54
5496	PKP *Ty2-507*	*+ 27.03.70*
5497	USSR (SW 48, Be 55, Li 58)	+ 01.01.72
5498	PKP *Ty2-508*	*+ 29.06.79*
5499	ÖStB, ÖBB 152.5499 (Amstetten 63)	+ 22.01.69
52 5500	PKP *Ty2-509*	*+ 30.04.87*
5501	PKP *Ty2-510*	*+ 11.06.87*
5502	DR (Berlin Schöneweide 70)	+ 02.06.76
5503	PKP *Ty2-511*	*+ 11.09.87*
5504	USSR (La 47, Ba 53, Es 56)	i 58
52 5505	USSR (La 45, Be 46, SW 48, i 56 Buntmetallindustrie)	
5506	PKP *Ty2-512*	*+ 15.01.77*
5507	USSR (Lw 53, Li 57), (63) PKP Ty2-1296	+ 19.10.90
5508	USSR 1042.374 (Krasnowka 93)	
5509	DR (Leipzig Süd 70)	+ 02.03.78
52 5510	PKP	
5511	SWDE, DB (Freiburg 50)	+ 18.10.54
5512	PKP *Ty2-514*	*+ 15.12.87*
5513	PKP *Ty2-515*	*+ 08.02.91*
5514		
5515	PKP *Ty2-516*	*+ 07.05.87*
5516	USSR (OK 53, Od 56, Be 61), (62) PKP Ty2-1224	+ 87
5517	PKP *Ty2-517*	*+ 20.02.73*
5518		
5519	USSR (Li 46, Ba 53, Li 56), (62) DR 52 8161 z 08.87 M	
52 5520	PKP *Ty2-518*	
5521	CCCPd	
5522	CSD 555.078, 555.3078	+ 66
5523	SWDE, DB (Ehrang 50)	+ 18.10.54
5524	DR (Berlin Schöneweide 70)	+ 25.08.78
5525	USSR (Kln 53, Mo 61), (64) BDZ 15.261	
5526	PKP *Ty2-519*	*+ 07.06.85*
5527	USSR 1042.375 (Bagrationowsk 93)	
5528	DB	+ 18.10.54
5529	USSR Uk 1042.376 (Rawa Russkaja 92)	
52 5530	USSR (Li 45 , Ba 53, Es 56)	i 63
5531	USSR (Kow50, Wnz 52, SW 53)	i 56
5532	USSR (SW48, Be 55, Li 58), → PKP Ty2-1293	+ 89
5533	USSR (Li 44)	
5534	USSR 1042.377 (Kaliningrad 93)	
5535	CCCPd PKP? *Ty2-1202*	*+ 05.10.90*
5536	PKP? *Ty2-520*	*+ 01.03.80*
5537	PKP *Ty2-1107*	*+ 19.07.79*
5538	USSR (OK 53, Mo 56, Be 59), (63) PKP	
5539	DB	+ 18.10.54
52 5540	USSR Be 1042.378 (Wysokoje 93)	
5541	PKP *Ty2-521*	*+ 15.06.82*
5542	DR (Altenburg 70)	+ 20.02.75
5543	USSR, (63) CSD 555.0298 555.3298	+ 67
5544	PKP *Ty2-1108*	
5545	DR(o)	+ 47
5546	DR (Dresden 70)	+ 02.06.75
5547	DB	+ 18.10.54
5548	USSR (Od 44-47)	
5549	DB	+ 18.10.54
52 5550	ÖStB	+ 47
5551	PKP *Ty2-522*	*+ 14.06.82*
5552	ÖStB, ÖBB	+ 20.11.54
5553	USSR (OK 53, Ba 55, La 56)	+ 01.05.75
5554	USSR (SW 49, S 53)	i 57
5555	DB	+ 18.10.54
5556	USSR (Be 53, Mo 60), (63) MAV 520,070	+ 03.67
5557	USSR (OK 53, Ba 54, Es 56)	i 57
5558	DB	+ 18.10.54
5559	USSR (SW 48, Be 55, La 58)	i 58
52 5560	USSR (Be 58) + 86 → Vietnam	x 94 ?
5561	USSR (Rezekne 59) 1042.380	+ 88
5562	USSR (Li 45, Be 47, SW 48, i 56 Kohlenindustrie)	
5563	PKP *Ty2-523*	*+ 06.10.77*
5564	PKP *Ty2-524*	*+ 25.11.89*
5565	PKP *Ty2-525*	
5566	USSR (Okt 53, Li 62), (63) PKP	Ty2-1361 ?
5567	PKP *Ty2-526*	*+ 04.11.89*
5568	(USSR La 47)? PKP *Ty2-527*	*+ 19.09.89*
5569	PKP *Ty2-528*	*+ 10.07.86*
52 5570	PKP *Ty2-529*	*+ 01.09.84*
5571	PKP *Ty2-530*	*+ 18.08.89*
5572	USSR Uk 1042.381 (Grodowiza 92)	
5573	NSB 5573	+ 20.08.64
5574	PKP *Ty2-531*	*+ 21.02.88*
5575	PKP *Ty2-532*	*+ 21.01.82*
5576	USSR (Ba53, La56, Lw59), (63) MAV 520,022	+ 03.67
5577	USSR (Königsberg Wrk 46)	
5578	DR (Leipzig Süd 70)	+ 20.07.78
5579	USSR (Kln 51, Okt 55, Mo 61), (64) BDZ	15.251 ?

52 5580	PKP	*Ty2-533*	
5581	PKP	*Ty2-534*	
5582	DB		+ 54
5583	USSR	1042.382 (Jelgava 85)	+ 89
5584	DR	52 8104 (Hoyerswerda 70)	Dsp 88 M
5585	DR	52 8121 (Eberswalde 70)	M
5586	JDZ	33-119	
5587	DB		+ 18.10.54
5588	SWDE, DB	(Friedrichshafen 50)	+ 18.10.54
5589	USSR	(Li 45, La 47, Kln 51, Okt 61)	+ 01.01.79
52 5590	USSR	(Od 50, Len 51, Okt 53, i 57 Kohlenindustrie)	
5591	PKP	*Ty2-535*	*+ 16.12.91*
5592	USSR	(SW 48, i 56, Do 61, i 61 Karakuba)	
5593	USSR	(Lw 46, Be 54), → Pi (Szczakowa 91)	
5594	USSR	(Radviliskis 54) 1042.383	+ 88
5595	USSR	(63) MAV 520,083 (09.63) GySEV	+ 08.81
5596	DR(w)		+ 48
5597	USSR	1042.384 (Krasnowka 93)	
5598	USSR	(SW53, La56, Lw59), (63) MAV 520,012	+ 03.67
5599	USSR	(Li 45, Lw 61)	
52 5600	PKP	*Ty2-536*	*+ 09.04.84*
5601	PKP	*Ty2-537*	
5602	PKP	*Ty2-538*	*+ 23.07.92*
5603	PKP	*Ty2-539*	*+ 16.11.85*
5604	USSR	(Kln 51, Li 59), (63) PKP	
5605	PKP	*Ty2-540*	
5606	NSB	5606 (Støren 70)	+ 04.70
5607	USSR	(Mo 60), (63) MAV 520,044	+ 10.71
5608	PKP	*Ty2-541*	*+ 05.02.74*
5609	USSR	(Li 45, Be 48, i 57 Kohlenindustrie)	
52 5610	USSR	Uk 1042.385 (Wjerbka 89, Rawa Russkaja 92)	
5611	USSR	(Ki 48, Lw 53, Li 57)	+ 01.11.75
5612	PKP	*Ty2-542*	*+ 19.12.73*
5613	USSR	(MK 48, Mo 59) (64) BDZ 15.240	
5614	PKP	*Ty2-543*	*+ 04.11.89*
5615	DB		+ 18.10.54
5616			
5617	USSR	(Preußisch Eylau 47)	+ 48
5618	USSR	(La 47, Ba 53, La 56)	+ 01.01.76
5619	USSR	(Be 50, Li 59), (63) PKP	
52 5620	USSR	1042.386 (Mamonowo 93)	
5621	USSR	(Li 45, Be 47, La 57)	i 76
5622	USSR	(Lw 53), (64) BDZ 15.226	
5623	PKP	*Ty2-544*	*+ 86*
5624	PKP	*Ty2-545*	*+ 86*
5625	ÖStB, ÖBB		+ 25.10.61
5626	DB		+ 18.10.54
5627	USSR	(Wjerbka 89)	+ 88
5628	USSR	(Be 53, Lw 61), (64) JZ 33-278	
5629	PKP	*Ty2-546*	*+ 07.09.88*
52 5630	PKP	*Ty2-547*	*+ 16.06.86*
5631	(CSD 555.0139) ?		
5632	PKP	*Ty2-548*	*+ 05.10.90*
5633	PKP	*Ty2-549*	*+ 15.12.87*
5634	CSD	555.0118 555.3118	+ 70
5635	SWDE, DB	(Radolfzell 50)	+ 18.10.54
5636	PKP	*Ty2-550*	*+ 11.02.86*
5637	USSR,	(62) CSD 555.0202 555.3202	+ 68
5638	USSR	(Ba 53, Es 56, Lw 57)	+ 01.04.75
5639	DR		+ 19.05.76
52 5640	PKP	*Ty2-551*	
5641	USSR	(Li 45, La 47, Ba 53, Es 56)	i 58
5642	DR	(Bautzen 70)	+ 06.06.77
5643	PKP	*Ty2-552*	*+ 04.02.75*
5644	PKP	*Ty2-553*	*+ 09.03.79*
5645	USSR	(Li 45, SW 48, La 56)	i 65
5646	PKP	*Ty2-554*	*+ 28.12.89*
5647	PKP	*Ty2-555*	*+ 20.07.77*
5648	PKP	*Ty2-556*	*+ 10.03.84*
5649	USSR	(La 46)	
52 5650	USSR	(MK53, SW59, Be61), (62) PKP Ty2-1225	+ 87
5651	DB		+ 18.10.54
5652	PKP	*Ty2-557*	*+ 31.08.87*
5653	USSR	Be 1042.388 (Osipowitschi 93)	
5654	PKP	*Ty2-558*	
52 5655	PKP	*Ty2-1191*	
5656	USSR	(Valga 56) 1042.389	+ 89
5657	PKP	*Ty2-559*	*+ 16.08.93*
5658	DR,	(59) BDZ 15.131 ?	
5659	DR	52 8079 (Frankfurt/Oder 70)	+ 07.95 M
52 5660	DR	(Wittenberg 70)	+ 05.04.90
5661	PKP	*Ty2-561*	*+ 29.05.85*
5662	KrV		+ 44
5663	DR		+ 13.01.69
5664	NSB	5664 (Krossen 62)	+ 24.09.63
5665	ÖStB, ÖBB		+ 20.11.54
5666	USSR	(OK 53, Od 56, i 57 Kohlenindustrie)	
5667	USSR	(OK 53, Od 56, Lw 62), (66) JZ 33-334	
5668	USSR	(MK 53, Be55, Mo60), BDZ 15.150	
5669	USSR	(Kln51, Li 57, Be58), (62) DR (Aue 70)	+19.06.75
52 5670	USSR	(Ba 53, La 56, Li 59), → PKP	
5671	USSR	(Tarnopol 46)	+ 46
5672	(USSR Lw 44)	PKP *Ty2-562*	*+ 31.12.87*
5673	PKP	*Ty2-563*	*+ 15.10.80*
5674	PKP	*Ty2-564*	*+ 02.12.90*
5675	USSR	(SW 51, Len 51, Okt 53, Be 58)	
		(62) DR (Hoyerswerda 70)	+ 11.10.76
5676	PKP	*Ty2-565*	*+ 15.07.67*
5677	PKP	*Ty2-566*	*+ 29.12.86*
5678	DR	52 8156 (Stendal 70)	
5679	DR	(Falkenberg 70)	z 88 M
52 5680	USSR	(Kln 53, Be 57), → Pi (Szczakowa 91)	M
5681	CCCPd		
5682	USSR	(Lw 48)	+ 55
5683	PKP		
5684	DB		+ 18.10.54
5685	DR	(Berlin Lichtenberg 70)	+ 04.01.77
5686	DR	(Leipzig West 70)	+ 29.05.81
5687	SNCF	150 Y 5687	
5688	ÖStB	KrV	+ 47
5689	USSR	Be 1042.390 (Wysokoje 93)	
52 5690	USSR	1042.391 (Zheleznodorozhnyj 93)	
5691	USSR	(Kln 53, Mo 59), (64) BDZ 15.179	
5692	USSR	(Li 45, SW 48, i 57 Kohlenindustrie)	
5693	USSR	Be 1042.392 (Neman 93)	
5694	USSR	(Rahmen ex 52 422)	
5695	PKP	*Ty2-568*	*+ 18.11.82*
5696	DR	(Frankfurt/Oder 70)	+ 22.11.78
5697	USSR	(OK 53, Mo 56), (60) BDZ	
5698	USSR	1042.618 (Jelgava 85)	+ 89
5699	PKP	*Ty2-569*	*+ 17.08.92*
52 5700	PKP	*Ty2-570*	*+ 14.11.73*
5701	PKP	*Ty2-571*	*+ 25.03.87*
5702	PKP	Ty2-572 (Bydgoszcz 96)	
5703	USSR	Uk 1042.394 (Wjerbka 89, Rawa Russkaja 92)	
5704	DR	(Stralsund 70)	+ 02.06.75
5705	USSR	1042.395 (Mamonowo 92)	
5706	FS, (49) DB		+ 18.10.54
5707	PKP	*Ty2-573*	*+ 20.02.88*
5708	FS, (49) DB		+ 18.10.54
5709	PKP	*Ty2-574*	*+ 26.07.89*
52 5710	USSR		+ 85
5711	PKP	*Ty2-575*	
5712	PKP	*Ty2-576*	*+ 20.09.89*
5713	DB, (51) JDZ		
5714	USSR	(Li 47)	i 51
5715	USSR	(Kln 53, Be 58, Li 58)	+ 01.05.76
5716	DB		+ 18.10.54
5717	USSR	Uk 1042.397	
5718	USSR	(OK 53, Od 56, Lw62), (64) JZ 33-305	
5719	USSR	(62) CSD 555.0201 555.3201	+ 69
52 5720	PKP	*Ty2-577*	*+ 14.01.87*
5721	PKP	*Ty2-578*	*+ 11.02.86*
5722	PKP	*Ty2-579*	*+ 20.12.82*
5723	DB		+ 18.10.54
5724	PKP	*Ty2-580*	*+ 11.09.74*
5725	USSR	(Ba 53, La 56, Mo 60), (64) BDZ 15.257	
5726	DB		+ 18.10.54
5727	DB		+ 18.10.54
5728	DR		+ 06.11.68
5729	ÖStB, ÖBB		+ 28.10.54

52 5730 CCCPd
5731 USSR 1042.398 (Zheleznodorozhnyj 93)
5732 USSR (La 56, Li 59), (62) DR (Zittau 70) + 24.10.78
5733
5734 PKP *Ty2-581* *+ 06.07.83*
5735 DB (München Ost 50) + 18.10.54
5736 PKP *Ty2-582*
5737 PKP *Ty2-583*
5738 USSR (1947)
5739 USSR (BL 45, Be 47, SW 49, i 56 Buntmetallindustrie)
52 5740 PKP *Ty2-584* *+ 27.08.88*
5741 PKP *Ty2-585* *+ 26.07.89*
5742 DR (Berlin Schöneweide 70) + 30.11.77
5743 PKP *Ty2-586* *+ 17.03.90*
5744 PKP *Ty2-587* *+ 25.03.77*
5745 PKP *Ty2-588* *+ 31.08.87*
5746 PKP *Ty2-589* *+ 08.10.76*
5747 PKP *Ty2-590* *+ 17.03.83*
5748 PKP *Ty2-591* *+ 16.04.85*
5749 DB, (52) JDZ 33-232
52 5750 USSR (Kln 51, Okt 55, Lw 61), (66) JZ 33-326
5751 PKP *Ty2-592* *+ 19.04.83*
5752 PKP *Ty2-593* *+ 15.01.88*
5753 USSR 1042.399 + 88
5754 PKP *Ty2-594* *+ 26.07.89*
5755 PKP *Ty2-1192* *+ 20.09.88*
5756 USSR (Okt 53, Lw 60) (64) JZ 33-279
5757 USSR (Kow 46, Wnz 52, SW 53) i 56
5758 USSR (SW 48, Od 52, OK 53, Mo 56, Lw 61)
(62) DR (Brandenburg 70) v 27.07.76
5759 PKP *Ty2-595* *+ 15.05.86*
52 5760 DB + 18.10.54
5761 DR 52 8035 (Falkenberg 70) z 87 M
5762 DR 52 9762 (Senftenberg 70) + 22.12.78
5763 DR (Frankfurt/Oder 70) + 03.07.70
5764 PKP *Ty2-596*
5765 PKP *Ty2-597* *+ 24.08.88*
5766 CCCPd
5767 DB + 18.10.54
5768 CCCPd
5769 USSR (Tapa 55) 1042.400 + 87
52 5770 USSR (Li 45, SW 48) i 56
5771 FS, (49) DB + 18.10.54
5772 USSR (Li 45, La 47, Ba 53, La 56) + 01.01.74
5773 USSR (Li 48)
5774 DB + 18.10.54
5775 DR (Leipzig West 70) + 26.09.69
5776 USSR (Li 47, Len 53, Okt 53) + 01.03.71
5777 USSR 1042.401 (Vilnius 85) + 88
5778 PKP *Ty2-1109* *+ 11.10.79*
5779 PKP *Ty2-598*
52 5780 USSR (Li 45, La 47, Ba 53, Es 56) + 01.06.74
5781 PKP *Ty2-599* *+ 19.11.91*
5782
5783 PKP *Ty2-600* *+ 15.09.76*
5784 CCCPö
5785 USSR (BL 45 , SW 49, S 53, i 56 Kohlenindustrie)
5786 PKP *Ty2-601* *+ 04.11.89*
5787 PKP *Ty2-602* *+ 03.05.86*
5788 USSR, (62) CSD 555.0244 555.3244 + 73
5789 DR 52 8045 (Falkenberg 70) + 15.08.88
52 5790 DB, (52) JDZ
5791 (USSR Lw 44) PKP *Ty2-603* *+ 01.09.84*
5792 PKP *Ty2-604* *+ 10.12.87*
5793 DR (Bautzen 70) + 06.02.78
5794 USSR (Lw 44) PKP? *Ty2-1205* *+ 15.01.77*
5795 ÖStB, ÖBB (Straßhof 63) + 28.04.76
5796 DB + 18.10.54
5797 PKP *Ty2- 1110* *+ 15.12.87*
5798 FS, (47) DB + 18.10.54
5799 DR (Pasewalk 70) + 17.03.70
52 5800 USSR (SW 48, Od 52, OK53) i 55
5801 PKP *Ty2-605*
5802 PKP *Ty2-606* *+ 30.06.71*
5803 DR 52 8137 (Stendal 70) z 86 M
5804 ÖStB, ÖBB (Knittelfeld 63) + 26.03.75 M
52 5805 DB + 18.10.54
5806 USSR Be 1042.402 (Wysokoje 93) M
5807 PKP *Ty2-1111* *+ 15.05.90*
5808 PKP
5809 PKP Ty2-1113 (Gniezno 92) + 11.04.92
52 5810 NSB 5810 + 24.09.63
5811 USSR (BL 50, Be 53, i 57 Kohlenindustrie)
5812 DR (Frankfurt/Oder 70) + 14.04.76
5813 USSR (Li 45, La 47, Kln 48, Ba 53, La 56)+ 01.02.73
5814 USSR (Ki 50, Len 52, Okt 53, i 57 Kohlenindustrie)
5815 DB, (52) JDZ 33-233
5816 PKP *Ty2-1114* *+ 17.03.90*
5817 DR (Dresden 70) + 28.10.77
5818 DB + 18.10.54
5819 USSR (SW 48, S 53, i 56 Kohlenindustrie)
52 5820 USSR (Preußisch Eylau 47) + 48
5821 PKP *Ty2-1115* *+ 19.04.91*
5822 PKP *Ty2-1116*
5823 USSR (Li 45)
5824 DB + 18.10.54
5825 USSR (Li 45, La 48, SW 48, Es 56) + 01.07.81
5826 USSR Be 1042.403 (Wysokoje 93)
5827 PKP *Ty2-607* *+ 01.08.85*
5828 PKP Ty2-1117 (Milkowice 91)
5829 USSR (SW 48, Ba 56, Li 56) + 01.04.71
52 5830 ÖStB, ÖBB (St Valentin 63) + 15.07.70
5831 USSR (OK 53, Ba 54, Es56, i 67 Kaliningrad)
5832 PKP *Ty2-1118*
5833 DB + 13.12.50
5834 PKP *Ty2-1119* *+ 04.11.89*
5835 PKP *Ty2-1120*
5836 PKP *Ty2-608* *+ 19.05.78*
5837 USSR, (63) CSD 555.0266 555.3266 + 68
5838 PKP *Ty2-1121* *+ 05.11.91*
5839 NSB 5839 (Trondheim 70) + 04.70
52 5840 USSR (Tschernjachowsk 85)
5841 NSB 5841 (Trondheim 70) + 11.70
5842 NSB 5842 + 24.09.63
5843 NSB 5843 (Støren 70) + 04.70
5844 NSB 5844 + 20.08.64
5845 NSB 5845 + 04.70
5846 NSB 5846 + 20.08.64
5847 NSB 5847 + 20.08.64
5848 NSB 5848 + 24.09.63
5849 NSB 5849 + 24.09.63
52 5850 NSB 5850 + 24.09.63
5851 NSB 5851 + 04.70
5852 NSB 5852 + 18.07.63
5853 NSB 5853 + 24.09.63
5854 NSB 5854 + 20.08.64
5855 NSB 5855 + 04.70
5856 NSB 5856 + 04.70
5857 NSB 5857 (Trondheim 60) + 04.70
5858 NSB 5858 + 20.08.64
5859 USSR (Li 47, Ba 53, La 56) + 11.04.64
52 5860 NSB 5860 + 04.70
5861 DB + 18.10.54
5862 NSB 5862 + 24.09.63
5863 NSB + 04.70
5864 DR (Frankfurt/Oder 70) + 04.01.77
5865 NSB + 04.70 M
5866 DB + 18.10.54
5867 USSR (BL 45, ... , La 50, Ba 53, La 56, Be 58)
(62) DR (Elsterwerda 70) + 24.02.76
5868 USSR (BL 47) i 51
5869 PKP *Ty2-1123*
52 5870 PKP
5871 DR (Angermünde 70) + 03.08.73
5872
5873 USSR (SW 48, S 53) i 56
5874 USSR Be 1042.405 (Brusgi 93)

52 5875 - 52 6343

SCHWARTZKOPFF (BMAG)

52 5875 - 6321 2/43 - 12/43
52 6322 - 6434 1/44 - 5/44

---

52 5875	USSR (Wnz 44)	
5876	DB	+ 18.10.54
5877	PKP	
5878	PKP *Ty2-610*	*+ 29.08.73*
5879	USSR (Od 50, Len 51, Okt 53)	i 73
52 5880	ÖStB, ÖBB 152.5880 (Amstetten 63)	+ 21.06.66
5881	PKP *Ty2-611*	*+ 04.11.89*
5882	USSR (BL 45, Be 46, SW 48)	i 56
5883	PKP *Ty2-612*	*+ 08.04.81*
5884	DR (Frankfurt/Oder 70)	+ 20.04.77
5885	DR (Dresden 70)	+ 11.10.76
5886	USSR 1042.406 (Zheleznodorozhnyj 93)	
5887	USSR (Okt 53, Li 62), (63) PKP Ty2-1357	+ 24.02.90
5888	PKP *Ty2-613*	
5889	DB (München Ost 50)	+ 18.10.54
52 5890	USSR 1042.407 (Krasnowka 93)	
5891	DB	+ 18.10.54
5892	USSR Be 1042.408 (Luninez 93)	
5893	USSR 1042.409 (Radviliskis 85)	+ 90
5894	PKP *Ty2-614*	*+ 15.01.88*
5895	USSR (Li 45, Ba 53, Es 56, Lw 57)	+ 01.10.71
5896	CCCPd	
5897	USSR Be 1042.410 (Wysokoje 93)	
5898	USSR (Od 44)	+ 48
5899	JDZ 33-120	
52 5900	DB	+ 18.10.54
5901	USSR (Od 44)	i 45
5902	DB	+ 18.10.54
5903	USSR (SW 48, Be 55), → Pi	
5904	ÖStB, ÖBB 152.5904 (Amstetten 63)	+ 06.05.67
5905	DR, (56) BDZ 15.105	
5906	USSR	i 75
5907	USSR 1042.411 (Vilnius 85)	+ 88
5908	USSR (BL 44)	
5909	USSR (50) → TE-8021	
52 5910	USSR (Od 44)	i 45
5911	USSR (Od 48, OK 53, Od 56)	i 74
5912		
5913		
5914	PKP *Ty2-615*	*+ 17.03.90*
5915	USSR (OK 53, Mo 56), (64) BDZ 15.215	M
5916	DR (Hoyerswerda 70)	+ 25.10.74
5917	USSR 1042.412 (Kaliningrad 93)	
5918	DR (Engelsdorf 70)	+ 02.03.78
5919	USSR Uk 1042.413 (Rawa Russkaja 92)	
52 5920	USSR (OK 53, Ba 55, La 56)	+ 01.04.68
5921	USSR (Jelgava 59) 1042.414	+ 88
5922	DB	+ 18.10.54
5923	USSR Uk (Lw 91) 1042.415 (Wjerbka 89)	
5924	USSR	
5925	USSR	M
5926	USSR 1042.417 (Tschernjachowsk 85)	+ 89
5927	USSR (Kln 53, Okt 61, i 73 Betonwerk Rshew)	
5928	JDZ 33-121	
5929	PKP Ty2-94 (Zabrzeg 94)	
52 5930	DB	+ 18.10.54
5931	DR (Frankfurt/Oder 70)	+ 26.09.77
5932	DR	+ 05.06.68
5933	USSR 1042.418 (Bagrationowsk 92)	M
5934	USSR (Wnz 44)	
5935	(USSR Lw 44) PKP *Ty2-616*	*+ 15.05.90*
5936	PKP *Ty2-617*	*+ 04.11.89*
5937	DB	+ 18.10.54
5938	USSR (Wnz 44, BL 47, Be 53, La 58, Lw 59)	i 65
5939	DB	+ 18.10.54
52 5940	USSR (Li 45, Ki 50, OK 53, Od 56, Li 58)	+ 01.06.71
5941	USSR (SW 48, S 53, SW 53, i 57 Kohlenindustrie)	
5942	PKP *Ty2-618*	*+ 23.02.90*
5943	USSR (Nowoseliza 46)	+ 48
5944	USSR (OK 53, Od 56, Li 58, Lw 62), (66) JZ 33-327	
5945	USSR Be 1042.419 (Wysokoje 93)	
5946	DB	+ 18.10.54
5947	PKP *Ty2-619*	*+ 01.06.87*
5948	USSR 1042.420 (Tschernjachowsk 85)	+ 86
5949	DB	+ 18.10.54
52 5950	USSR 1042.421 (Daugavpils 85)	+ 88
5951	USSR 1042.422 (Tschernjachowsk 85)	+ 88
5952	SWDE, DB (Freudenstadt 50)	+ 18.10.54
5953	USSR (Okt 53, i 57 Tambow)	+ 60
5954		
5955	USSR (IF 46, Kow 46, Lw 53)	+ 01.08.69
5956	SWDE, DB (Betzdorf 50)	+ 18.10.54
5957	PKP Ty2-620 (Jaworzyna 83)	+ 25.10.85
5958	PKP Ty2-869 (Lebork 93)	+ 15.05.91
5959		
52 5960	DR (Pasewalk 70)	v 17.12.76
5961	BDZ 15.55	
5962	PKP *Ty2-622*	*+ 10.05.76*
5963	JDZ 33-196	
5964	USSR Uk 1042.423 (Wjerbka 89)	
5965	USSR 1042.424 (Tschernjachowsk 93)	
5966	USSR (63) CSD 555.0299 555.3299	+ 67
5967	USSR (Tschernjachowsk 85) 1042.425	
5968	USSR (63) CSD 555.0285 555.3285	+ 67
5969	PKP *Ty2-623*	*+ 05.12.85*
52 5970	JDZ 33-122	
5971	FS, (49) DB	+ 18.10.54
5972	PKP *Ty2-624*	*+ 21.03.73*
5973	USSR Be 1042.426 (Wysokoje 93)	
5974	PKP *Ty2-625*	*+ 15.12.92*
5975	CSD ?	
5976	DB	+ 18.10.54
5977	DB	+ 18.10.54
5978	USSR + 86 → Vietnam (Hanoi 87)	x 94 ?
5979	USSR (Od 50, Len 51, Okt 53, i 67 Waldindustrie)	
52 5980	PKP *Ty2-626*	*+ 08.10.82*
5981	USSR (63) CSD 555.0291 555.3291	+ 71
5982	USSR Be 1042.428 (Wysokoje 93)	
5983		
5984	DR(w) KrV	+ 46
5985	JDZ 33-123	
5986	USSR (Be 45), (63) PKP Ty2-1339	+ 07.03.85
5987	PKP *Ty2-627*	
5988	DB	+ 18.10.54
5989	PKP *Ty2-628*	*+ 30.07.91*
52 5990	USSR (Kln51, Okt61, Mo62), (64) BDZ	
5991	PKP *Ty2-629*	*+ 12.09.88*
5992	DR 52 8111 (Görlitz 70)	z 87 M
5993	USSR (Od 44)	i 45
5994	USSR (Be 58), (64) JZ 33-306	
5995	ÖStB, ÖBB (Knittelfeld 63)	+ 21.03.75
5996	DR 52 8163 (Stendal 70)	M
5997	ÖStB	+ 47
5998	USSR (SW 48)	i 56
5999	DB	+ 20.01.52
52 6000	JDZ 33-124	
6001	DR (Falkenberg 70)	+ 30.07.75
6002	USSR (Lw 58), (63) MAV 520,092	+ 05.78
6003	USSR Be 1042.429 (Luninez 93)	
6004	USSR 1042.430 (Krasnowka 93)	
6005	DB	+ 18.10.54
6006	USSR (Mo 61), (63) MAV 520,005	+ 11.72
6007	USSR Ungeny 94	
6008		
6009	(23.08.44 Rumänien)	
52 6010	USSR (BL45, Li 48, Ba 53, Li 56)	i 64
6011	DR 52 8064 (Lübbenau 70)	z 87 M
6012	DB	+ 18.10.54
6013	USSR (Be 45, La 47, Ba 53, La 56)	i 72
6014		

52 6015	DB	+ 18.10.54
6016	DR (Berlin Pankow 69)	+ 22.02.79
6017	PKP *Ty2-630*	*+ 30.04.87*
6018	DB	+ 18.10.54
6019	USSR	
52 6020	USSR	+ 87
6021	USSR (Be 50, i 57 Kohle, Do 61, i 61)	
6022	DR (Stralsund 70)	+ 19.05.76
6023	PKP *Ty2-631*	*+ 17.03.90*
6024	FS, (47) DB	+ 18.10.54
6025	USSR (SW 53, La 56, Lw 59)	
6026	USSR (Lw 48)	+ 01.07.84
6027	PKP Ty2-632 (Andrzejowka 91)	+ 05.07.91
6028	DB	+ 18.10.54
6029	USSR, (62) PKP Ty2-1234 (Gniezno 93)	
52 6030	USSR (SW 44, Be 45, Ki 49)	i 51
6031	PKP *Ty2-633*	*+ 18.08.89*
6032	JDZ 33-125	
6033	USSR (Li 47, Ba 53, Li 56, i 73 Zementwerk Akmene)	
6034	SWDE, DB (Betzdorf 50)	+ 18.10.54
6035	SWDE, DB (Worms 50)	+ 18.10.54
6036	USSR Uk (Wjerbka 89)	
6037	PKP *Ty2-634*	
6038	PKP *Ty2-635*	
6039	DR(o) KrV	+ 47
52 6040	USSR (Tapa 85) 1042.435	+ 88
6041	USSR (Kln 51, Okt 61)	+ 01.08.82
6042	PKP *Ty2-636*	
6043	USSR Uk 1042.436 (Wjerbka 89, Rawa Russkaja 92)	
6044	USSR Uk	
52 6045	USSR (Od 49, OK 53, Od 56, i 57 Kohlenindustrie)	
6046		
6047	PKP Ty2-637 (CN 91)	+ 25.08.89
6048	(USSR Lw 44) PKP *Ty2-638*	*+ 15.09.76*
6049	DR (Berlin Schöneweide 70)	+ 16.05.77
52 6050	USSR (Kow 46)	i 52
6051	USSR (OK 53,Od 56,Mo58), BDZ 15.174 (Asenowo 90)	
6052	DB	+ 18.10.54
6053	USSR	+ 88
6054	USSR (Be 47), (63) PKP Ty2-1328	+ 19.11.92
6055	JDZ 33-126	
6056	DR (Pasewalk 70)	+ 01.02.77
6057	USSR Be 1042.439 (Wysokoje 93)	
6058	USSR (Be 53, Li 59), → PKP Ty2-1285	+ 23.07.92
6059	PKP *Ty2-639*	*+ 09.09.87*
52 6060	USSR Uk (Wjerbka 89)	
6061	USSR (Od 45)	
6062	TCDD 56526 (Kars 87)	
6063	TCDD 56527 (Balikesir 84)	
6064	PKP *Ty2-640*	*+ 16.04.86*
6065	ÖStB, ÖBB (Wien Ost 63)	+ 22.11.72
6066	TCDD 56528 (Afyon 81, Balikesir 92)	
6067	TCDD 56529 (Usak 81, Afyon 95)	
6068	TCDD 56530 (Afyon 81, Sivas Wrk 87)	
6069	TCDD 56531 (Izmir 81, Halkapinar Wrk 88)	

52 6070	TCDD	56532 (Erzurum 81, Kars 87)	
6071	TCDD	56533 (Afyon 75, Afyon 95)	
6072	TCDD	56534 (Eregli 75, Alasehir 95)	
6073	TCDD	56535 (Balikesir 75, Dinar 95)	
6074	CSD	555.096	
6075	USSR	(Kow 46, Lw 53)	+ 63
6076	USSR	Uk 1042.441 (Rawa Russkaja 92)	
6077	DR	(Falkenberg 70)	+ 19.05.76
6078	USSR	(St 44, Kow 45) PKP *Ty2-641*	*+ 15.05.86*
6079	DB		+ 13.12.50
52 6080	PKP		
6081	NSB	6081	+ 24.09.63
6082	DR	52 8152 (Görlitz 70)	M
6083	USSR	(Kln 52, Okt 55, Be 60), (62) PKP Ty2-1226	
6084	ÖStB, ÖBB	(Straßhof 63)	+ 28.04.77
6085			
6086	USSR	(62) CSD 555.0208 555.3208	+ 67
6087	USSR	(Len 51, Okt 53, Es 56)	i 66
6088	USSR	(Be 50, Kln 54, Okt 61)	+ 01.07.75
6089	USSR	(NW45, Be46, SW 59, Mo 62), (64) BDZ 15.265	
52 6090	USSR	(Lw 47)	
6091	PKP	*Ty2-643*	*+ 18.11.86*
6092	DB		+ 18.10.54
6093	DR(w)	KrV	+ 20.02.47
6094	PKP	Ty2-644 (Stalowa Wola 91)	
6095	ÖStB,	ÖBB + 23.12.57 → Ungarn	
6096	USSR	Be 1042.610 (Berestowiza 93)	
6097	USSR	(St 44)	
6098	USSR	(Od 44)	
6099	PKP	*Ty2-645*	*+ 14.06.89*
52 6100	USSR	(Be 50, i 57 Kohlenindustrie)	
6101	PKP	*Ty2-646*	*+ 17.03.83*
6102	PKP	*Ty2-647*	*+ 11.09.87*
6103	PKP	*Ty2-648*	*+ 30.04.87*
6104	PKP	*Ty2-649*	*+ 15.03.84*
6105	USSR	1042.442 (Kaliningrad 92)	+ 90
6106	DR	52 8095 (Frankfurt/Oder 70)	M
6107	PKP?	*Ty2-650*	*+ 09.08.90*
6108	USSR	(Od 44)	
6109	JDZ	33-127	
52 6110	USSR	Uk 1042.443 (Rawa Russkaja 92)	
6111	DB		+ 18.10.54
6112	USSR	(Lw 46, Ki 48)	i 51
6113	USSR	(Li 45-50)	
6114	USSR	(Od 44)	
6115	USSR	Be 1042.444 (Luninez 93)	M
6116	USSR	(IF46, Od 48, OK53, Mo 56, Lw 61), (64) JZ	
6117	USSR	(Be 58) Uk	
6118	DR(o)	KrV	+ 47
6119	USSR	(Wnz 44, ..., Okt 55, Li 59), → Ty2-1277	+ 86
52 6120	USSR	(Len 51, Okt 53, Li 59), (64) JZ 33-318	
6121	USSR	(Wnz 44)	+ 46
6122	USSR	(63) CSD 555.0258 555.3258	+ 70
6123	PKP	*Ty2-651*	*+ 17.08.83*
6124	USSR		
6125	DR	(Frankfurt/Oder 70)	+ 19.05.76
6126	DB	(52) JDZ 33-257	
6127	DR(o)		+ 12.50
6128	USSR,	(63) MAV 520,071 #	+ 08.74
6129	USSR	(OK 53, Od 56, Mo 58), Export	
52 6130	SWDE, DB	(Betzdorf 50)	+ 18.10.54
6131	USSR	(Wnz 52, SW 53, i 56 Kohlenindustrie)	+ 61
6132	USSR	(Od 44)	
6133	USSR	Rahmen → TE-293	
6134	ÖStB, ÖBB	(St Veit/Glan 63)	+ 29.11.77
6135	SWDE, DB	(Worms 50)	+ 18.10.54
6136	DB		+ 18.10.54
6137	USSR	Uk 1042.447 (Grodowiza 92)	
6138	USSR	(SW 44)	
6139	USSR		+ 88
52 6140	USSR	(Wnz 45)	
6141	JDZ	33-228	
6142	USSR	(Li 47, Ba 53, La 56, Lw 61), (66) JZ 33-325	
6143	PKP	Ty2-1098 (Koscierzyna 90)	
6144	DR	(Karl-Marx-Stadt 70)	+ 12.07.76
52 6145	PKP		
6146	USSR	(Okt 53, Lw 61), (64) JZ 33-281	
6147	PKP	*Ty2-654*	*+ 16.06.88*
6148	USSR	(OK 53, Od 56, Be 58), (62) PKP Ty2-1207	+ 88
6149	USSR	(Wnz 44) Tender: Uk Rawa Russkaja 92	
52 6150	USSR	(Wnz 44)	+ 46
6151	USSR	1042.449 (Bagrationowsk 93)	+ 91
6152	PKP	*Ty2-655*	*+ 26.10.78*
6153	DB		+ 18.10.54
6154	PKP	*Ty2-656*	*+ 18.12.85*
6155	DB		+ 18.10.54
6156	DB		+ 18.10.54
6157	DR(w)	KrV	+ 06.03.46
6158	DR(w)	KrV + 46 oder PKP *Ty2-657*	*+ 29.11.90*
6159	DR	52 8106 (Falkenberg 70)	z 07.88 M
52 6160	USSR	(Riga 58) 1042.450	+ 87
6161	SWDE, DB		+ 18.10.54
6162	USSR	(BL 47, Be 53), (63) PKP Ty2-1327	+ 12.03.86
6163	FS, (51) DB		+ 18.10.54
6164	USSR	1042.451 (Kaliningrad 93)	
6165	USSR	(SW 50)	i 56
6166	USSR	(Od 44)	i 45
6167	USSR	(Od 56, Mo 60), (63) MAV 520,033	+ 11.72
6168	USSR	(Wnz 44)	
6169	USSR	(BL 50, Be 53, Li 58), (63) PKP Ty2-1270	+ 81
52 6170	PKP	*Ty2-658*	*+ 10.02.87*
6171	CCCPc		
6172	DR	(Kamenz 70)	+ 23.04.75
6173	USSR	1042.452 (Kaliningrad 93)	+ 91
6174	DB		+ 18.10.54
6175	DB		+ 18.10.54
6176	USSR	(Riga 58) 1042.453	+ 90
6177	USSR	(Wnz 45)	+ 46
6178	SNCF	150 Y 6178	
6179	USSR	Uk 1042.454 (Rawa Russkaja 92)	
52 6180	PKP	*Ty2-659*	*+ 05.10.90*
6181	PKP	*Ty2-660*	*+ 15.05.90*
6182	PKP	*Ty2-661*	
6183	DR	(Elsterwerda 70)	+ 09.12.76
6184	ÖStB	KrV	+ 01.03.46
6185	DB		+ 18.10.54
6186	DB		+ 18.10.54
6187	PKP	*Ty2-662*	*+ 27.07.87*
6188	PKP	*Ty2-663*	*+ 19.10.90*
6189	DR(w)	KrV	+ 20.09.48
52 6190	USSR	1042.623 (Krasnowka 93)	
6191	PKP	Ty2-1035 (Jaworzyna 91)	M
6192	PKP	*Ty2-665*	*+ 21.07.83*
6193			
6194	DR	(Frankfurt/Oder 70)	+ 29.10.69
6195	DB		+ 18.10.54
6196	USSR	(Be 46, Lw 62), (66) JZ 33-338	
6197	DB		+ 18.10.54
6198	DB		+ 18.10.54
6199	DR	(Dresden 70)	+ 16.01.75
52 6200	USSR	(BL45, Be 48), (63) PKP Ty2-1245	+ 23.02.90
6201	USSR	(Lw 47), (64) BDZ	
6202	ÖStB,	ÖBB	+ 20.11.54
6203	USSR	(Lw 46, BL 47, Be 53, Lw 61)	+ 01.09.80
6204	NSB	6204 (Oslo 62)	+ 20.08.64
6205	PKP	*Ty2-666*	*+ 23.07.92*
6206	DR	(Riesa 70)	+ 20.07.78
6207	USSR	(Li 46, Be 47, St 52, SW53, i 56 Kohlenindustrie)	
6208	PKP	Ty2-1129	+ 25.09.92
6209	USSR	Uk	
52 6210	SWDE, DB	(Betzdorf 50)	+ 18.10.54
6211	DB		+ 18.10.54
6212	USSR	(i)	
6213	JDZ	33-128	
6214	PKP	*Ty2-668*	*+ 15.01.77*
6215	PKP	*Ty2-669*	*+ 15.09.89*
6216	DB		+ 18.10.54
6217	DR	(Hoyerswerda 70)	+ 19.05.76
6218	PKP	*Ty2-670*	
6219	PKP	*Ty2-88*	*+ 17.03.90*

52 6220 USSR (Ki 44) + 51
6221 DR (Kamenz 70) + 01.02.77
6222 USSR Uk
6223 USSR 1042.457 (Bagrationowsk 93)
6224
6225 DR (Güsten 70) + 25.10.74
6226 DR + 11.02.53
6227 DR (Frankfurt/Oder 70) + 29.07.76
6228 PKP *Ty2-671*
6229 DR (Hoyerswerda 70) + 29.07.76
52 6230 DB + 18.10.54
6231 PKP
6232 PKP *Ty2-673*
6233 PKP *Ty2-674* *+ 01.12.81*
6234
6235 CSD
6236 SWDE, DB (Ehrang 50) + 18.10.54
6237 USSR
6238 FS, (48) DB, (52) JDZ 33-259
6239 USSR, (64) BDZ 15.270 (Kalojanowetz 92)
52 6240 PKP *Ty2-675* *+ 19.11.91*
6241 USSR (62) CSD 555.0220 555.3220 + 71
6242 DB (52) JDZ 33-247
6243 PKP *Ty2-676* *+ 12.08.76*
6244 USSR (Lw 44) → PKP *Ty2-677* *+ 12.06.73*
6245 DB (52) JDZ 33-241
6246 DB + 18.10.54
6247 USSR (Lw 44) → PKP *Ty2-678* *+ 28.04.87*
6248 JDZ 33-129
6249 PKP *Ty2-679* *+ 05.07.91*
52 6250 PKP *Ty2-680* *+ 19.05.83*
6251 JDZ 33-130
6252 DB + 18.10.54
6253 USSR (Lw 53, Mo 60), (63) MAV 520,034 + 11.72
6254 BDZ 15.56 (Asenowo 92)
6255 PKP *Ty2-681* *+ 15.01.88*
6256 USSR (Kow 50, Lw 53, La 57) + 01.05.70
6257 DB + 18.10.54
6258 USSR (Ki 52, Lw 53) + 01.07.65
6259 JDZ 33-197
52 6260 USSR (SW 48, Ba 61), (63) PKP Ty2-1360 + 08.09.76
6261 USSR (La 50, Kln 51, Li 58),
(62) DR (Angermünde 70) + 04.08.69
6262 PKP *Ty2-682* *+ 16.06.89*
6263 FS, (49) DB, (52) JDZ 33-263
6264 PKP *Ty2-683* *+ 28.11.80*
6265 DR (Görlitz 70) + 06.09.75
6266 NSB 6266 + 04.70
6267 PKP *Ty2-684* *+ 10.04.92*
6268
6269 DR (Pasewalk 70) + 29.09.80
52 6270 USSR Uk
6271 PKP *Ty2-685 → Pi (KWK Rozbark 91)*
6272 DB + 18.10.54
6273 PKP Ty2-137 (Tczew 90) + 28.04.90
6274 USSR (Be 58), Uk (Wjerbka 89)
6275 PKP *Ty2-687* *+ 01.09.84*
6276 PKP
6277 DR 52 8164 (Brandenburg 70) + 15.08.88
6278 DB + 18.10.54
6279 ÖStB, ÖBB + 30.04.48
52 6280 DB + 18.10.54
6281 (Tender → USSR, Uk Rawa Russkaja 92)
6282 DB + 18.10.54
6283 USSR (Kow 46, Lw 53, La 58, i 75 Zuckerfabrik Kursk)
6284 DB + 18.10.54
6285 USSR (La 47, Li 59) + 01.06.70
6286 PKP *Ty2-689* *+ 22.07.91*
6287 USSR (BL 47, Be 53, Mo 60), (64) BDZ 15.235
6288 ÖStB, ÖBB (Wels 63) + 28.04.77
6289 USSR (MK 53, Kln 58, Li 59), (63) PKP
52 6290 FS (48) DB + 20.05.49
6291 USSR (Li 47) i 62
6292 USSR (Od 44, Ki 48) Export
6293 PKP Ty2-690 (Chelm 92) + 03.11.89
6294 PKP *Ty2-691* *+ 10.11.73*

52 6295 USSR (Wnz 44)
6296 PKP *Ty2-692* *+ 11.02.86*
6297 DB + 18.10.54
6298 DB, (51) JDZ 33-235
6299 PKP *Ty2-693*
52 6300 DR 52 8125 (Hoyerswerda 70) z 07.87 M
6301 USSR (Li 45, Ki 48, Be 60), (63) PKP
6302 USSR (Lw 53), (63) MAV 520,023 + 11.72
6303 USSR (Li 50, La 56, Okt 61, i 74 Tallinn)
6304 USSR (MK 53, Be55, Lw 62), (64) JZ (33-307?)
6305 PKP *Ty2-694* *+ 16.08.76*
6306 DR (Zittau 70) + 06.02.78
6307 DR(o) + 46
6308 ÖStB, ÖBB + 20.11.54
6309 PKP *Ty2-695*
52 6310 DB + 18.10.54
6311 DR (Berlin Schöneweide 70) + 15.03.73
6312 ÖStB, ÖBB (Knittelfeld 63) + 15.04.75 § 985.223
6313 DR (Zittau 70) + 14.11.75
6314 DR (Kamenz 70) + 23.04.75
6315 USSR (Be 50, MK 54, Lw 58) + 01.05.69
6316 DB, (51) JDZ 33-245
6317 DB + 13.12.50
6318 PKP *Ty2-696* *+ 89*
6319 USSR (Wnz 44, SW 48) →TE-8022?
(63) → PKP Ty2-1311 (Tomaszow 91) + 90
52 6320 DR (Pasewalk 70) + 14.04.76
6321 PKP *Ty2-697* *+ 29.11.80*
6322 PKP Ty2-775 (Gniezno 93)
6323 ÖStB + 09.02.47
6324 USSR (63) CSD 555.0275 555.3275 + 68
6325 USSR (Lw 53), (64) BDZ 15.201 (Kalojanowetz 92)
6326 USSR (Li 44)
6327 PKP *Ty2-699* *+ 24.12.86*
6328 SWDE, DB (Mainz 50) + 18.10.54
6329 USSR (Li 45)
52 6330 USSR (La 45, Li 47) + 01.12.76
6331 PKP *Ty2-1199* *+ 15.10.76*
6332
6333 PKP? *Ty2-700* *+ 07.05.74*
6334 USSR (Li 45, Od 48, i 56 Kohlenindustrie)
6335 PKP *Ty2-701* *+ 22.11.83*
6336 DR 52 8022 (Frankfurt / Oder 70) + 18.03.83
6337 DR (Salzwedel 70) + 20.08.76
6338 DB + 18.10.54
6339 DR 52 8193 (Templin 70) + 19.08.88
52 6340 PKP Ty2-702 (Zbaszynek 92) + 05.06.93
6341 DB (Worms 50) + 18.10.54
6342 DB + 18.10.54 oder BDZ 15.57 ?
6343 USSR (Li 45) ?
6344 USSR (Li 58), (62) DR (Wittenberg 70) + 06.02.78
6345 PKP *Ty2-1125* *+ 04.04.85*
6346 USSR 1042.461 (Krasnowka 93)
6347 SWDE, DB + 18.10.54
6348 USSR (Ki 44, ... , Li 59),
(62) DR (Wittenberg 70) + 05.11.76
6349 BDZ 15.58
52 6350
6351 USSR (Ki 45, Lw61), (62) DR(Salzwedel 70) + 16.09.77
6352 USSR (SW 48) i 56
6353 USSR (La 47, Es 59, Okt 63) i 73
6354 PKP *Ty2-1126* *+ 21.01.87*
6355 DR (Frankfurt/Oder 70) + 23.01.79
6356 DR (Stralsund 70) v 75
6357 DR 52 8003 (Eberswalde 70) z 12.87
6358 DR (Berlin Schöneweide 70) + 21.02.77
6359 DR 52 8047 (Cottbus 70) M
52 6360 USSR (Kow 45, Ki 48, Li 58) i 58
6361 CSD 555.0124, (59) BDZ 15.124 (Asenowo 92)
6362 CSD
6363 CSD 555.083 555.3083 + 67
6364 CSD 555.0122 + 71
6365 USSR Be 1042.462 (Lida 93) M
6366 CSD 555.084 + 68
6367 CSD 555.0126 555.3126 + 70
6368 CSD
6369 CSD

52 6370	CSD	555.089, (59) → BDZ 15.115 (Kalojanowetz 92)	
6371	DB		+ 18.10.54
6372	DR	52 8066 (Frankfurt/Oder 70)	+ 09.08.88
6373	DR	(Wustermark 70)	x 09.81
6374	USSR	(La 47)	+ 01.12.73
6375	USSR	(BL 45, Li 47)	+ 01.01.73
6376	USSR	(Od 50, Mo 60), (64) BDZ 15.214 (Asenowo 90)	
6377	CCCPc	(CSD 555.0140) ?	
6378	PKP	*Ty2-1127*	
6379			
52 6380	SWDE, DB	(Freudenstadt 50)	+ 18.10.54
6381	PKP	*Ty2-1128*	*+ 01.06.82*
6382	USSR	1042.463 (Jelgava 85)	+ 89
6383	CCCPc	(CSD 555.0141) ?	
6384	PKP		
6385	USSR	1042.464 (Vilnius 85)	+ 89
6386	USSR	(Be 53, Mo 60), (63) MAV 520,035	+ 03.68
6387	USSR	(Ba 56, Be 58), (63) PKP	
6388	USSR	(Be 58)	
6389	PKP	Ty2-1177 (Nowy Sacz 88)	+ 16.09.91
52 6390	DR	52 8122 (Eberswalde 70)	M
6391	USSR	(Li 45)	
6392	USSR	(Li 46, Be 48, Li 59, Mo 60), BDZ 15.153	
6393	PKP	*Ty2-1130*	*+ 17.03.90*
6394	PKP	*Ty2-1131*	*+ 28.12.89*
6395	PKP	Ty2-709 (Zagan 91)	+ 16.12.91
6396	NSB	6396	+ 04.70
6397	USSR	(Be 50, Lw 61), (64) JZ 33-293	
6398	DR	(Falkenberg 70)	+ 30.07.75
6399	USSR	1042.466 (Kaliningrad 85)	+ 89
52 6400	DR		+ 27.10.71
6401	PKP	*Ty2-1133*	*+ 16 04.85*
6402	USSR	(Be 47, Li 59), (64) JZ 33-321	
6403	PKP	Ty2-703 (Lukow 93)	+ 05.02.91
6404	DR	(Altenburg 70)	+ 18.11.85
6405	USSR	(Li 47)	+ 01.01.72
6406	DB		+ 18.10.54
6407	ÖStB, ÖBB	(Mürzzuschlag 63)	+ 12.11.75
6408	PKP	*Ty2-1134*	
6409	PKP	*Ty2-1135*	*+ 29.11.90*
52 6410	USSR	(La 47, Be 58), (63) PKP Ty2-1353	+ 92
6411	DR, (59) BDZ	(15.133 ?)	
6412	NSB	6412	+ 24.09.63
6413	USSR	(Li 45, La 48)	+ 01.01.81
6414	PKP	*Ty2-1136*	
6415	DR	52 8065 (Frankfurt/Oder 70)	+ 11.04.74
6416	PKP	*Ty2-1137*	*+ 12.06.78*
6417	DR		+ 10.50
6418	PKP	*Ty2-1138*	
6419	USSR	(Be 47, Li 57)	+ 01.11.71
52 6420	DR	(Brandenburg 70)	v 27.06.77
6421	DR	(Zwickau 70)	+ 06.09.75
6422	USSR	(Be 47, MK 53, Be 55, SW 59)	i 60
6423	USSR	Be 1042.467 (Wysokoje 93)	
6424	DB		+ 18.10.54
6425	BDZ	15.59? Rahmen → 15.85 (Todor Kableschkow 90)	
6426	BDZ	15.60 (Bodrowo 92)	
6427	DR	52 8074 (Frankfurt/Oder 70)	+ 19.08.88
6428	ÖStB, ÖBB	(Straßhof 63) + 27.10.78	§ 9760 034
6429	PKP	*Ty2-1139*	*+ 04.11.89*
52 6430	DB		+ 18.10.54
6431	DR	(Karl-Marx-Stadt 70)	+ 19.08.77
6432	USSR	Be 1042.468 (Wysokoje 93)	
6433	USSR	Be 1042.469 (Wysokoje 93)	
6434	DB		+ 18.10.54

52 6625 - 52 6684

SKODA

5/43 - 11/43

---

52 6625	CSD	555.001 555.3001	+ 67
6626	PKP	*Ty2-704*	*+ 17.04.91*
6627			
6628	PKP	*Ty2-705*	*+ 04.11.89*
6629	DB		+ 10.08.57
52 6630	DB		+ 10.08.57
6631	DR, (56) BDZ	15.90 (Asenowo 92)	
6632	DR	(Frankfurt/Oder 70)	+ 01.09.80
6633	PKP		
6634	DR	(Hoyerswerda 70)	+ 17.03.76
6635	CSD	555.002	+ 69
6636	PKP	*Ty2-707*	*+ 04.11.89*
6637	PKP	*Ty2-708*	*+ 17.03.90*
6638	DB		+ 10.08.57
6639	DR	(Seddin 70)	+ 29.02.72
52 6640	PKP		
6641	USSR	(Lw 46, Ki 48, Wnz 51)	i 56
6642	USSR	(Li 47)	+ 01.06.81
6643	PKP	*Ty2-710*	*+ 05.10.90*
6644	USSR	(Ostaschkow 88)	+ 05.88 M
6645	CSD	555.003 555.3003	+ 71
6646			
6647	DR(o)		+ 46
6648	USSR, (63) CSD	555.0292	
6649	ÖStB, ÖBB	+ 10.08.59 § 985.211	
52 6650	CSD	555.004, (59) BDZ 15.113 (Asenowo 92)	
6651	DB		+ 15.11.57
6652	PKP	*Ty2-711*	
6653	CSD	555.057 555.3057	+ 69
6654	PKP	*Ty2-712*	*+ 07.06.69*
6655	CSD	555.005 ?	
6656	PKP	*Ty2-713*	
6657	DR	(Elsterwerda 70)	+ 20.02.75
6658	DR	(Stralsund 70)	+ 03.08.73
6659	DR	(Dresden 70)	+ 19.12.74
52 6660	DR	(Dresden 70)	+ 19.06.75
6661	CSD	555.006 (52) 555.0609	
		(56) 555.0179 555.3179	+ 68
6662	USSR	(S 53, Be 55), (63) PKP Ty2-1348	
6663	CSD	555.007 555.3007	+ 71
6664	ÖStB, ÖBB		+ 25.01.48
6665	PKP	Ty2-390 (Walbrzych 94)	+ 23.10.92
6666	DR	(Jüterbog 70)	M
6667	USSR	1042.471 1042.617	+ 88
6668	USSR	(BL 45, Li 47)	+ 01.02.71
6669	USSR	(La 47)	+ 06.62
52 6670	CSD	555.009 555.3009	+ 68
6671	CSD	555.010	+ 72
6672	CSD	555.011 555.3011	+ 73
6673	PKP	*Ty2-715* oder CSD 555.064 ?	
6674	CSD	555.012	+ 71
6675	USSR	(Be 58), Ungeny 94	
6676	USSR	(BL 47, Lw 48)	+ 01.09.81
6677	USSR	(Kln 53, Be 57), (62) PKP	
6678	USSR	(La 47)	+ 01.04.71
6679	PKP	*Ty2-716*	*+ 15.01.77*
52 6680	PKP	*Ty2-717*	*+ 14.06.89*
6681	USSR	(Be 47, MK 53, Li 58)	i 58
6682	PKP	*Ty2-718*	*+ 16.02.77*
6683	USSR	1042.473 (Bagrationowsk 93)	
6684	DR	52 8021 (Frankfurt/Oder 70)	+ 07.95 M

52 6685 - 52 7434

Wiener Lokomotivfabrik

52 6685 - 7427 2/43 - 12/43
52 7428 - 7434 1/44

---

52 6685	DB		+ 18.10.54
6686	JDZ	33-198	
6687	JDZ	33-131	
6688	DB	(Treuchtlingen 50)	+ 18.10.54
6689	ÖStB, ÖBB	(Graz 63)	+ 22.11.72
52 6690	USSR	→ Pi Ty2-6690	
6691	DB		+ 21.04.52
6692	USSR	(Be 50, Kln 54, Okt 61)	i 65
6693	USSR	(Be 50, SW 59, i 60 Kiew Wagen Reparaturwerk)	
6694	DR	(56) BDZ 15.104	
6695	USSR	Be 1042.474 (Berestowiza 93)	
6696	USSR		
6697	PKP	*Ty2-719*	*+ 06.02.91*
6698	CCCPd		
6699	USSR	(Li 45, La 48, SW49, Lw 49, Od 52)	+ 81
52 6700	DR	(Pasewalk 70)	+ 02.01.69
6701	DR	(Seddin 70)	+ 13.10.80
6702	USSR		
6703	PKP	*Ty2-720*	*+ 20.09.88*
6704	PKP	*Ty2-721*	*+ 09.08.90*
6705	PKP	*Ty2-722*	*+ 23.04.75*
6706	DR	(Seddin 70)	+ 14.05.79
6707	USSR	(Li 45)	
6708	DR	(Kamenz 70)	+ 29.07.76
6709	SWDE, DB	(Worms 50)	+ 18.10.54
52 6710	PKP	*Ty2-723*	*+ 25.09.91*
6711	PKP	*Ty2-724*	*+ 04.11.89*
6712	DR	(Stralsund 70)	+ 24.10.78
6713	FS, (47) DB		+ 21.04.52
6714	USSR	Be 1042.611 (Wysokoje 93)	
6715	ÖStB, ÖBB	152.6715	+ 06.03.69
6716	PKP	*Ty2-725*	*+ 11.02.86*
6717	ÖStB ÖBB	152.6717	+ 20.11.68
6718	JDZ	33-199	
6719	USSR	Be 1042.476 (Wysokoje 93)	
52 6720	USSR	(Len 51, Okt 53, i 57 Kohlenindustrie)	
6721	DR	(Cottbus 80)	z 08.86 M
6722	USSR	Be 1042.477 (Berestowiza 93)	
6723	DB		+ 21.04.52
6724	DB		+ 18.10.54
6725	USSR	1042.619 (Bagrationowsk 93) #	
6726	PKP	*Ty2-726*	*+ 25.06.74*
6727	PKP		
6728	PKP	*Ty2-728*	*+ 13.09.88*
6729	ÖStB, ÖBB	152.6729 (70) GKB	+ 75
52 6730	USSR	(La 47)	+ 01.06.73
6731	PKP	*Ty2-729*	*+ 02.06.87*
6732	USSR	Be 1042.478 (Wysokoje 93)	
6733	USSR	(Kow 46, La 57)	+ 01.08.73
6734	USSR	(SW 44)	
6735	PKP	*Ty2-730*	*+ 26.10.85*
6736	DB		+ 18.10.54
6737	USSR	(La 47)	+ 01.11.75
6738	PKP?	*Ty2-731*	*+ 24.08.85*
6739	DR(w)		+ 12.07.48
52 6740	DR	(Berlin Schöneweide 70)	+ 06.06.77
6741	PKP	*Ty2-732*	*+ 19.05.78*
6742	PKP	*Ty2-733*	*+ 29.11.90*
6743	DB		+ 18.10.54
6744	USSR	1042.479 (Gusew 92)	M
6745	BDZ	15.61	
6746	USSR	(Kow 48, La 57, Kln 58, Okt 61)	+ 01.04.80
6747	USSR	1042.480 (Riga 92)	x 92
6748			
6749	USSR	1042.481 (Daugavpils 85)	+ 89

52 6750	USSR	(NK46, Kln51, Okt 55, Li 62)	+ 88
6751			
6752	USSR	1042.482 (Valga 89)	+ 89
6753	SWDE, DB	(Kirn 50)	+ 18.10.54
6754	PKP	*Ty2-734*	*+ 05.07.84*
6755			
6756	USSR	(Od 44)	+ 55
6757	DB		+ 18.10.54
6758	USSR	Uk	
6759	USSR	(Li 45)	+ 51
52 6760	USSR	(Ki 47, La 58)	+ 01.09.76
6761	USSR	1042.484 (Riga 92)	
6762	USSR	(Lw 46, Ki 47, La 57)	+ 01.04.71
6763	USSR	(Tschernjachowsk 85) 1042.485	+ 88
6764	PKP	*Ty2-735*	
6765	ÖStB, ÖBB	(Knittelfeld 63) + 20.12.68	§ 985.218
6766	ÖStB, ÖBB	(Selzthal 63)	+ 15.07.70
6767	USSR	(BL 47)	i 51
6768	USSR	(Luninez 90)	
6769	USSR	1042.487 (Luninez 90)	M
52 6770	USSR	(Lw 53), (63) CSSR	
6771	USSR, TE-6771	→ Pi Ty2-6771 (Szczakowa 91) TE-6771" 1042 488 (Radviliskis 85)	
6772	USSR	(Kln 51, Okt 55, Li 58), (63) PKP	
6773	USSR	(Kln 51, Be 58), (62) PKP	
6774	DR	(Templin 70)	+ 04.01.77
6775	USSR	(Riga 60)	
6776	DB		+ 18.10.54
6777	USSR	Li 57 Unbestätigt	
6778	DR	52 8056 (Bautzen 70)	z 07.87 M
6779	DR	52 8051 (Elsterwerda 70)	M
52 6780	PKP	*Ty2-736*	
6781	PKP	*Ty2-737*	*+ 21.07.76*
6782	USSR	Be 1042.490 (Brusgi 93)	
6783	PKP	*Ty2-738*	*+ 24.12.86*
6784	PKP	*Ty2-739*	*+ 27.06.72*
6785	PKP	*Ty2-740*	*+ 04.11.89*
6786	PKP	*Ty2-741*	*+ 16.06.86*
6787	PKP	*Ty2-742*	
6788	USSR	(Lw 59), (63) MAV 520,024	+ 08.65
6789	PKP	*Ty2-743*	*+ 23.02.90*
52 6790	PKP	*Ty2-744*	*+ 14.07.86*
6791	USSR	(Li 45, Ki 48, Lw 53), (64) BDZ 15.221	
6792	PKP	*Ty2-745*	
6793	PKP	*Ty2-746*	*+ 08.07.86*
6794	ÖStB, ÖBB	(Selzthal 63)	+ 27.10.78
6795	JDZ	33-132	
6796	DB		+ 18.10.54
6797	USSR	(Li 50, Be 58), (62) PKP Ty2-1212	+ 20.11.87
6798	USSR	1042.491 (Zheleznodorozhnyj 93)	
6799	USSR	(Be 50), (62) PKP Ty2-1215	
52 6800	USSR	(Len 51, Lw 61), (64) JZ 33-294	
6801	BDZ	15.62 (Bodrowo 92)	
6802	USSR	(Be 58) Uk	
6803	USSR	(Lw 46, Od 48)	i 56
6804	USSR	(Okt 53, Be 58), (63) PKP Ty2-1322 → Pi	
6805			
6806	ÖStB, ÖBB	(Selzthal 63)	+ 22.11.72
6807	DB		+ 18.10.54
6808	CCCPö		
6809			
52 6810	ÖStB, ÖBB		+ 15.03.56
6811	DR	(Berlin Schöneweide 70)	+ 25.08.78
6812	USSR	(BL 50), (63) PKP Ty2-1319	+ 27.05.81
6813	USSR	(SW 50, Len 51, i 57 Waldindustrie)	
6814	USSR, (62) CSD	555.0231 555.3231	+ 69
6815	DB		+ 18.10.54
6816	DR	52 8107 (Lübbenau 70)	+ 19.08.88
6817	USSR	(Kow 45, Li 58, Lw 63)	+ 01.09.72
6818	USSR, (62) CSD	555.0210 555.3210	+ 68
6819	ÖStB, ÖBB	(Knittelfeld 63)	+ 27.10.78
52 6820	USSR	(Li 47)	i 64
6821	DR	(Engelsdorf 70)	+ 19.08.75
6822	USSR	Be 1042.493 (Wysokoje 93)	
6823	CSD	555.0142 555.3142	+ 71
6824	USSR, (62) CSD	555.0249 555.3249	+ 70

52 6825	PKP *Ty2-748*	
6826	USSR (La 47, Kln 51, i 56 Kohlenindustrie)	
6827	PKP *Ty2-749*	*+ 28.04.87*
6828	DR (Neubrandenburg 70)	+ 02.06.75
6829	DR 52 8157 (Stendal 70)	z 04.87 M
52 6830	DR (Frankfurt/Oder 70)	+ 30.10.69
6831	USSR (Kln 54, Mo 61), (64) BDZ 15.236 oder 15.232	
6832	USSR 1042.494 (Luninez 90)	M
6833	USSR, (63) CSD 555.0263 555.3263	+ 71
6834	USSR Be 1042.495 (Wysokoje 93)	
6835	USSR (Li 47)	+ 01.08.74
6836	DR (Cottbus 70)	+ 14.07.75
6837	CSD	
6838	USSR (Radviliskis 55) 1042.496	
6839	PKP *Ty2-750*	*+ 20.11.87*
52 6840	USSR (Be 45, SW 49, i 57 Kohlenindustrie)	
6841	ÖStB, ÖBB (Straßhof 63)	+ 12.11.75
6842	PKP *Ty2-751*	
6843	DB	+ 18.10.54
6844	JDZ 33-133	
6845	ÖStB, ÖBB (Straßhof 63)	+ 27.10.78
6846	JDZ 33-200	
6847	FS (49) DB	+ 18.10.54
6848	SNCF 150 Y 6848	+ 18.07.55
6849	DB	+ 18.10.54
52 6850	CCCPm (CFR 150.1111)?	
6851	USSR (SW 50)	i 56
6852	PKP *Ty2-752*	*+ 08.04.71*
6853	USSR (Od 51, Ba 53, La 56)	+ 01.05.74
6854	USSR (Li 49, Mo 59), (64) BDZ 15.196 (Kalojanowetz 92)	
6855	USSR (Lw 46, La 48, SW 48, S 53)	i 56
6856	CCCPc	
6857	USSR (Lw 46, Ki 48), Export	
6858	USSR (Be 58, Ba 85) 1042.497	+ 88
6859	ÖStB, ÖBB (Wien Ost 63)	+ 01.08.70
52 6860	DR (Güsten 70)	+ 05.06.68
6861	USSR (SW 48, Be 55), (62) PKP Ty2-1210	
6862	USSR (Lw 46, Li 47)	i 75
6863	USSR (BL 50, Kln 54), (63) PKP	
6864	USSR (SW 47, S 53)	i 56
6865	DR	+ 02.51
6866	DB	+ 18.10.54
6867	USSR (SW 50)	i 56
6868	USSR 1042.498 (Mamonowo 92)	+ 91
6869	PKP *Ty2-753*	
52 6870	PKP *Ty2-754*	*+ 04.10.74*
6871	PKP Ty2-755 (Jaworzyna 84)	+ 10.11.87
6872	PKP *Ty2-756*	*+ 04.06.76*
6873	DR 52 8146 (Jüterbog 70)	+ 15.08.88
6874	PKP *Ty2-757*	*+ 30.08.75*
6875	PKP *Ty2-758*	
6876	DB	+ 18.10.54
6877	SWDE, DB (Friedrichshafen 50)	+ 18.10.54
6878	PKP *Ty2-759*	*+ 15.05.90*
6879	USSR (Li 50, La 56, Lw 62), (66) JZ 33-332	
52 6880	ÖStB, ÖBB (Graz 63)	+ 20.03.70
6881	USSR (Len 51, Okt 53, i 57 Kohlenindustrie)	
6882	USSR (Ki 51, Ba 55, La 56)	+ 01.12.76
6883	USSR, (62) PKP Ty2-1269 (Olsztyn 92)	+ 25.09.92
6884	SWDE, DB (Mainz 50)	+ 18.10.54
6885		
6886	DB	+ 18.10.54
6887	DB	+ 18.10.54
6888	USSR (Kln 51, Okt 61)	+ 01.08.71
6889	JDZ 33-201	
52 6890	JDZ 33-202	
6891	ÖStB, ÖBB (Straßhof 63)	+ 15.06.73
6892	USSR (Kaliningrad 85) 1042.499	+ 88
6893	ÖStB, ÖBB (Straßhof 63)	+ 13.12.73
6894	ÖStB	+ 01.09.46
6895	ÖStB	+ 01.09.46
6896	CCCPö	
6897	CSD	
6898	ÖStB	+ 15.04.46
6899	ÖStB, ÖBB (Straßhof 63)	+ 16.12.74

52 6900	ÖStB, ÖBB (Graz 63)	+ 16.06.75
6901	DR (Pasewalk 70)	+ 15.03.73
6902	DR (Karl-Marx-Stadt 70)	+ 22.07.77
6903	DR	+ 03.06.54
6904	DR (Dresden 70)	+ 06.02.78
6905	PKP *Ty2-760*	*+ 06.10.75*
6906	PKP Ty2-761 (Jaworzyna 87)	+ 10.11.87
6907	PKP Ty2-762 (Jaworzyna 87)	+ 10.11.87
6908	PKP *Ty2-763*	*+ 11.04.92*
6909	USSR (62) CSD 555.0230 555.3230	+ 68
52 6910	USSR (Be 58, Ba 85) 1042.500	+ 89
6911	DB	+ 18.10.54
6912	DR (Kamenz 70)	+ 04.01.78
6913	USSR, (63) PKP Ty2-1383	+ 09.03.85
6914	PKP Ty2-967 (Lebork 93)	+ 02.12.90
6915	USSR (62) CSD 555.0219 555.3219	+ 69
6916	USSR	
6917	USSR (Lw 46, Ki 48, Li 59), (64) JZ 33-319	
6918	PKP Ty2-765 (Nysa 84)	
6919	USSR Uk (Rawa Russkaja 92)	
52 6920	USSR (Li 45, Len 51, Lw 61)	
	(62) DR (Frankfurt/Oder 70)	+ 01.07.77
6921	USSR Be 1042.503 (Wysokoje 93)	
6922	DR (Berlin Schöneweide 70)	+ 16.05.77
6923	USSR (Od 44, NW45, Lw 45, Be 46, Li 57)	+ 01.05.75
6924	USSR (La 47)	+ 01.04.72
6925	PKP Ty2-891 (Lebork 93)	+ 04.11.89
6926	PKP *Ty2-767*	*+ 16.01.76*
6927	USSR (62) CSD 555.0215 555.3215	+ 67
6928	DR (Zwickau 70)	+ 20.02.75
6929	PKP *Ty2-768*	*+ 28.12.89*
52 6930	DR (Zittau 70)	v 14.09.76
6931	SWDE, DB (Radolfzell 50)	+ 18.10.54
6932	DR (Kamenz 70)	+ 14.04.76
6933	JDZ 33-134	
6934	SAAR, (57) DB	+ 20.07.59
6935		
6936	DB	+ 18.10.54
6937	USSR (Ki 51, Be 59), (63) PKP Ty2-1251	
6938	ÖStB, ÖBB (Wien Ost 63)	+ 20.01.73
6939	USSR (Lw 46, Ki 48, Wnz 51)	i 56
52 6940	ÖStB, ÖBB + 15.05.60	§ 985.213
6941	ÖStB, ÖBB (Straßhof 63)	+ 26.03.73
6942	USSR (Kln 51, Lw 59), (66) JZ 33-331	
6943	USSR (Lw 46, Len 51, i 57 Kohlenindustrie)	
6944	USSR (Lw53) '63 MAV 520,079 (9/63) → GySEV	+ 6.80
ex52 4369	USSR TE-6944" (Tschernjachowsk 85) 1042.504	+ 89
6945	USSR (SW 52)	i 56
6946	USSR (La 50, Kln 51, Okt 61)	+ 01.04.83
6947	ÖStB, ÖBB (Wiener Neustadt 63)	+ 28.04.77
6948	CCCPm	
6949	USSR (SW 48, Be 55), (63) PKP Ty2-1310	+ 92
52 6950	USSR (Ki 51), BDZ 15.154 (Asenowo 92)	
6951	ÖStB, ÖBB (Mürzzuschlag 63)	+ 17.02.75
6952	USSR Be 1042.505 (Wysokoje 93)	
6953	USSR (Od 48, Mo 58), (64) BDZ 15.216 (Kalo'wetz 92)	
6954	USSR (Be 58), (63) PKP	
6955	USSR 1042.506 (Radviliskis 85)	+ 92
6956	JDZ 33-217 ?	
6957	USSR (63) CSD 555.0290 555.3290	+ 71
6958	JDZ 33-203	
6959	PKP *Ty2-769*	*+ 03.11.89*
52 6960	USSR (Tschernjachowsk 85) 1042.507	+ 91
6961	USSR (SW 50, Len 51, Okt 53)	i 57
6962	USSR (Daugavpils 54) 1042.508	+ 89
6963	USSR (Be 50, Sw 59, Lw 61),	
	(62) DR (Güsten 70)	v 23.08.75
6964	ÖStB, ÖBB (Knittelfeld 63)	+ 25.04.73
6965	USSR (Be 53)	i 57
6966	ÖStB, ÖBB (Linz 63)	+ 04.11.74
6967	ÖStB, ÖBB (Wien Ost 63)	+ 27.10.78
6968	ÖStB, ÖBB (Linz 63)	+ 16.06.75
6969	ÖStB, ÖBB (Straßhof 63) + 20.03.70	§ 985.219
52 6970	ÖStB, ÖBB (Linz 63)	+ 29.12.75
6971	USSR Uk (Wjerbka 89)	
6972	ÖStB, ÖBB (Straßhof 63)	+ 28.04.77
6973	JDZ 33-135	
6974	DB	+ 18.10.54

52 6975	FS, (49) DB	+ 18.10.54
6976	ÖStB, ÖBB (Graz 63)	+ 16.12.74
6977	DB	+ 21.04.52
6978	FS, (49) DB	+ 18.10.54
6979	DB	+ 18.10.54
52 6980	ÖStB, ÖBB (Wien Ost 63)	+ 01.02.73
6981	DB	+ 18.10.54
6982	SNCF 150 Y 6982	
6983	DB	+ 18.10.54
6984	DB	+ 18.10.54
6985	DB	+ 18.10.54
6986	DB	+ 18.10.54
6987	DR(w)	+ 21.05.47
6988	DB	+ 18.10.54
6989	DB	+ 21.04.52
52 6990	DB	+ 21.04.52
6991	DB	+ 18.10.54
6992	DR(w) Kesselexplosion	+ 12.11.47
6993	DB	+ 18.10.54
6994	USSR (Ki 45, Ba 54, Be 58), (63) PKP Ty2-1356	
6995	PKP Ty2-770 (Tczew 93)	+ 16.08.93
6996	DR(w) 10/45 → DR(o)→ DR 52 8078	x 08.93
6997	DB	+ 18.10.54
6998	DB	+ 21.04.52
6999	DB	+ 21.04.52
52 7000	DB	+ 18.10.54
7001	PKP Ty2-771 (Milkowice 91)	+ 15.12.88
7002	PKP *Ty2-772*	*+ 15.01.77*
7003	DR 52 8067 (Wustermark 70)	+ 03.08.73
7004	PKP *Ty2-773*	*+ 28.09.73*
7005	USSR, (64) BDZ 15 232 oder 15.236	
7006	USSR 1042.510 (Tschernjachowsk 85)	+ 89
7007	USSR (BL 50, Be 53, i 57 Kohlenindustrie)	
7008	PKP *Ty2-774*	*+ 10.11.87*
7009	DR (Kamenz 70)	+ 23.04.75
52 7010	USSR (Lw 53), (64) JZ 33-314	
7011	CCCPm	
7012	ÖStB, ÖBB (Graz 63)	+ 28.05.75
7013	USSR 1042.511 (Vilnius 85)	+ 86
7014	USSR 1042.512 (Daugavpils 85)	+ 92
7015	ÖStB, ÖBB (Graz 63)	+ 07.08.69
7016	KrV	+ 04.04.45
7017	ÖStB, ÖBB (Knittelfeld 63)	+ 28.04.77
7018	ÖStB, ÖBB (Graz 63)	+ 16.08.76
7019	JDZ 33-136	
52 7020	FS, (48) DB	+ 18.10.54
7021	USSR (Be 50, Kln 54, Be 58), (63) PKP	
7022	CCCPm	
7023	USSR, (63) PKP Ty2-1334 (Korsze 93)	+ 15.06.92
7024	DB	+ 18.10.54
7025	FS, (47) DB	+ 21.04.52
7026	DB	+ 21.04.52
7027	DB	+ 18.10.54
7028	USSR 1042.513 (Tschernjachowsk 85)	+ 91
7029	CCCPö	
52 7030	USSR, Uk (Lwow 89)	
7031	ÖStB, ÖBB (Knittelfeld 63)	+ 15.07.70
7032	USSR (Ki 51), BDZ 15.172	
7033	JDZ 33-137	
7034	ÖStB, ÖBB (Knittelfeld 63)	+ 15.07.70
7035	JDZ 33-138	
7036	JDZ 33-139	
7037	JDZ 33-140	
7038	JDZ 33-141	
7039	JDZ 33-142	
52 7040	USSR (SW 50)	i 56
7041	JDZ 33-143	
7042	USSR (Lw 46, BL 47), (62) PKP Ty2-1232	+ 17.07.87
7043	USSR, Be 1042.515 (Brest-Wostok 93)	
7044	USSR (Len 51)	i 56
7045	USSR, (63) PKP Ty2-1306 (Jaworzyna 87)	+ 17.03.90
7046	ÖStB, ÖBB (Selzthal 63)	+ 10.78 M
7047	USSR, (62) CSD 555.0221 555.3221	+ 72 M
7048	USSR (Lw 53), (64) BDZ 15.184	(Kalojanowetz 92)
7049	USSR (Len 51)	i 56
52 7050	CCCPc	
7051	ÖStB, ÖBB (Wiener Neustadt 63)	+ 27.02.75
7052	ÖStB, ÖBB (Knittelfeld 63) + 20.03.70	§ 985.220
7053	ÖStB, ÖBB (Knittelfeld 63) + 25.07.77	§ 9760 033
7054	USSR	
7055	JDZ 33-144	
7056	ÖStB, ÖBB	+ 15.06.62
7057	USSR, Uk (Rawa Russkaja 92) #	
7058	USSR (Len 51, La 57, Be 58), (62) DR (Dresden 70)	+ 06.09.75
7059	DR(w)	+ 20.09.48
52 7060	USSR (BL 50, Be 58), → Pi	
7061	ÖStB, ÖBB (Straßhof 63)	+ 28.04.77
7062	ÖStB, ÖBB (Straßhof 63)	+ 20.11.68
7063	JDZ 33-229	
7064	ÖStB, ÖBB (Selzthal 63)	+ 22.11.72
7065	ÖStB, ÖBB (Knittelfeld 63)	+ 01.02.73
7066	USSR, (62) CSD 555.0226 555.3226	+ 66
7067	ÖStB, ÖBB (Wien Ost 63)	+ 27.02.75
7068	USSR (Len 51, i 57 Kohlenindustrie)	
7069	FS, (49) DB	+ 18.10.54
52 7070	ÖStB, ÖBB (Knittelfeld 63)	+ 27.10.78
7071	DB	+ 18.10.54
7072	USSR (Kln 51, Mo 61), (63) MAV 520,006	+ 11.72
7073	ÖStB, ÖBB (Wien Ost 63)	+ 28.05.75
7074	USSR 1042.518 (Mamonowo 93)	
7075	USSR (Ki 50, Kln 55, Li 59), (64) JZ 33-322	
7076	JDZ 33-145	
7077	PKP	
7078	CCCPc (Zvolen 47)	
7079	USSR 1042.519 (Zheleznodorozhnyj 93)	
52 7080	PKP Ty2-1057 (Jaworzyna 91)	+ 29.04.91
7081	USSR (Ba 54, Lw 62), (66) JZ 33-341	
7082	USSR (Li 58), (63) PKP Ty2-1306	+ 17.03.90
7083	USSR (Be 58), (63) PKP Ty2-1331	+ 15.01.77
7084	ÖStB, ÖBB (Wien Ost 63)	+ 20.10.75
7085	ÖStB, ÖBB (Straßhof 63)	+ 20.10.71
7086	ÖStB KrV	+ 30.04.48
7087	ÖStB, ÖBB (Wien Ost 63)	+ 30.07.73
7088	ÖStB, ÖBB (Wien Ost 63)	+ 22.11.72
7089	PKP *Ty2-777*	*+ 23.02.90*
52 7090	PKP? *Ty2-834*	*+ 04.07.91*
7091	PKP Ty2-820 (Jarocin 89)	+ 07.10.92
7092	ÖStB, ÖBB (Straßhof 63)	+ 25.04.73
7093	USSR (Ki 51, Lw 62), (64) JZ 33-308	
7094	CCCPö	
7095	JDZ 33-204	
7096	USSR (Kln 51, Okt 55)	i 73
7097	USSR, Uk 1042.520 (Grodowiza 92)	
7098	USSR (Len 51)	+ 01.06.71
7099	JDZ 33-205	
52 7100	ÖStB, ÖBB (Graz 63) + 25.10.77	§ 985.230
7101	PKP *Ty2-779*	*+ 05.04.83*
7102	ÖStB, ÖBB (Wiener Neustadt 63)	+ 10.78 M
7103	DR	+ 27.12.71
7104	DR (Frankfurt/Oder 70)	+ 23.01.79
7105	PKP *Ty2-780*	*+ 10.02.87*
7106	SWDE DB (Friedrichshafen 50)	+ 18.10.54
7107	PKP *Ty2-781*	*+ 02.12.90*
7108	PKP *Ty2-782*	*+ 09.09.87*
7109	DR 52 8108 (Halle-G 70)	+ 28.05.91
52 7110	PKP *Ty2-783*	
7111	PKP *Ty2-784*	*+ 16.01.76*
7112	JDZ 33-206	
7113	USSR (La 50, Kln 51, Li 59), (62) DR	+ 70
7114	JDZ 33-146	
7115	USSR (Ki 51, Len 52, i 57 Kohlenindustrie)	
7116	ÖStB, ÖBB (Mürzzuschlag 63)	+ 27.10.78
7117	ÖStB, ÖBB (Graz 63)	+ 27.10.78
7118	USSR (Kln 51, Okt 61)	+ 01.07.84
7119	PKP *Ty2-785*	*+ 11.09.87*
52 7120	ÖStB, ÖBB (Wien Ost 63)	+ 20.10.72
7121	USSR (Kow 46, Li 58)	+ 01.09.78
7122	DR 52 8042 (Elsterwerda 70)	z 09.87 M
7123	USSR, Be 1042.521 (Neman 93)	
7124	DB	+ 18.10.54

52 7125	USSR (i)	
7126	DB	+ 18.10.54
7127	USSR (La 47)	+ 01.05.69
7128	USSR (Kln 51, Okt 61)	+ 01.04.71
7129	DR (Elsterwerda 70)	+ 23.04.75
52 7130	PKP *Ty2-786*	*+ 07.03.85*
7131	DR (Frankfurt/Oder 70)	+ 19.05.76
7132	DR (Eberswalde 70)	+ 19.05.76
7133	USSR, Uk 1042.522	
7134	USSR, Uk (Wjerbka 89)	
7135	DB	+ 18.10.54
7136	DR (Stralsund 70)	+ 19.12.74
7137		
7138	DR 52 8134 (Hoyerswerda 70)	+ 07.95 M
7139	DR (Pasewalk 70) #	+ 27.12.71
52 7140		
7141	CSD 555.0144, 555.3144	+ 68
7142	USSR (62) CSD 555.0245 555.3245	+ 68
7143	PKP *Ty2-787*	*+ 02.12.90*
7144	CCCPc (CSD 555.0145) ?	
7145	PKP *Ty2-788*	
7146	CSD 555.071	+ 70
7147	PKP *Ty2-789*	*+ 03.11.75*
7148	CCCPc (CSD 555.0146) ?	
7149	USSR (Okt 55, Mo 61), (63) MAV 520,054	+ 12.73
52 7150	CSD 555.0147 555.3147	+ 68
7151	PKP Ty2-197 (CN 91)	+ 18.06.86
7152	CCCPc (CSD 555.0148) ?	
7153	CCCPc (CSD 555.0149) ?	
7154	USSR (Kow 50, Lw 59)	
7155	USSR (SW 50, Be 62), → Pi (Szczakowa 91)	
7156	USSR (Kow 50, Wnz 52)	i 56
7157	USSR, Uk (Wjerbka 89)	
7158	USSR (Li 50, W 51, ... , Lw 61) (62) DR (Brandenburg 70)	+ 19.05.76
7159	USSR (Li 50, i 58)	+ 01.08.74
52 7160	USSR (Od 50, Lw 58), (63) MAV 520,093	+ 03.77
7161	USSR (Be 58)	
7162	USSR (Kln51,Li 59), (62) DR (Neustrelitz 70)	+ 03.05.74
7163	CSD	
7164	USSR, Uk	
7165	USSR (SW 50, Li 57, OK 64)	+ 01.01.78
7166		
7167	DB ?	
7168	USSR 1042.527 (Radviliskis 91)	x 91
7169	DB	+ 18.10.54
52 7170	USSR 1042.528 (Vilnius 85)	
7171	USSR 1042.529 (Mamonowo 93)	
7172	DR 52 8010 (Lübbenau 70)	M
7173	USSR (SW 50, Od 57, Be 62), → Pi (Szczakowa 90)	M
7174	PKP ? *Ty2-791* ?	
7175	USSR 1042.530 (Kaliningrad 85)	M
7176	SWDE, DB (Worms 50) USSR TE-7176" (Ungeny 94)	+ 18.10.54
7177	JDZ 33-147	
7178	DR (Berlin-Schöneweide 70)	+ 07.08.79
7179	PKP *Ty2-792*	*+ 03.11.89*
52 7180	USSR (Lw 45)	
7181	USSR (Lw 46, Ki 47)	i 55
7182	DB	+ 18.10.54
7183	CSD 555.0150 555.3150	+ 68
7184	PKP *Ty2-793*	
7185	USSR 1042.532 (Lwow-Zapad.93)	+ 88
7186	USSR (Len 51, Mo 60), (64) JZ 33-300	
7187	CSD 555.0101	+ 69
7188	USSR (Be 47)	i 53
7189	DR (Dresden 70)	+ 17.03.76
52 7190	USSR (La 57, Be 60), (62) PKP Ty2-1218	
7191	DR (Hoyerswerda 70)	+ 29.07.76
7192	PKP *Ty2-794*	*+ 06.04.74*
7193	JDZ 33-148	
7194	PKP *Ty2-795*	*+ 12.03.86*
7195	DR 52 9195 (Senftenberg 70)	+ 20.12.79
7196	USSR (i) Puksoosero	
7197	PKP *Ty2-796*	*+ 17.03.90*
7198	USSR, (62) CSD 555.0241 555.3241	+ 69
7199	USSR (Mo 61), (64) BDZ 15.266	
52 7200	PKP *Ty2-797*	
7201	USSR (Len 51, i 57 Kohlenindustrie)	
7202	JDZ 33-149	
7203	JDZ 33-150	
7204	CCCPö	
7205	USSR (Okt 55, Be 60), (62) PKP Ty2-1221	+ 12.03.91
7206	ÖStB, ÖBB (St.Valentin 63)	+ 15.07.70
7207	ÖStB, ÖBB (Graz 63)	+ 01.02.73
7208	USSR (Ki 51, Len 52, i 57 Kohlenindustrie)	
7209	DB	+ 18.10.54
52 7210	ÖStB, ÖBB (Wien Ost 63)	+ 15.06.73
7211	USSR (Len 51, i 57 Kohlenindustrie)	
7212	USSR (Ki 50, Kln 55, Be 59), (63) PKP Ty2-1316	+ 76
7213	ÖStB, ÖBB (Straßhof 63)	+ 25.10.77 § 985.228
7214	ÖStB, ÖBB (Mürzzuschlag 63)	+ 01.08.70
7215	ÖStB, ÖBB (Linz 63)	+ 20.10.72
7216	DR (Pasewalk 70)	+ 17.03.70
7217	PKP *Ty2-798*	
7218	PKP *Ty2-799*	
7219	USSR (SW 47, i 57 Kohlenindustrie)	
52 7220		
7221	DR 52 8117 (Frankfurt/Oder 70)	M
7222	DR (Berlin Lichtenberg 70)	+ 23.01.79
7223	DR 52 8091 (Berlin Schöneweide 70)	M
7224	USSR (Li 47)	i 64
7225	PKP *Ty2-800*	*+ 11.02.86*
7226	USSR (SW 51)	+ 63
7227	DR (Frankfurt/Oder 70)	+ 20.04.77
7228	PKP *Ty2-801*	*+ 19.04.78*
7229	USSR 1042.533 (Tapa 85)	+ 90
52 7230	PKP *Ty2-802*	*+ 25.09.88*
7231	USSR 1042.534 (Daugavpils 85)	+ 90
7232	USSR (Be 58)	
7233	USSR (La 47)	+ 01.06.76
7234	USSR 1042.535 (Tapa 85)	+ 88
7235	PKP *Ty2-803*	
7236		
7237	USSR (Len 51, Li 59), (64) JZ 33-320	
7238	DR	+ 27.10.69
7239	JDZ 33-151	
52 7240	PKP *Ty2-804*	*+ 20.11.87*
7241	DR	+ 17.03.76
7242	DR 52 8150 (Berlin Schöneweide 70)	M
7243	DB	+ 18.10.54
7244	PKP *Ty2-805*	*+ 13.10.82*
7245	DB	+ 18.10.54
7246	DR (Cottbus 70)	+ 23.04.75
7247	DR 52 8027 (Falkenberg 70)	+ 06.03.91
7248	DR 52 8057 (Görlitz 70)	z 12.86 M
7249	PKP *Ty2-1197*	*+ 12.02.86*
52 7250	USSR 1042.536 (Mamonowo 93)	
7251	DR 52 8167 (Brandenburg 70)	+ 15.08.88
7252	DR (Leipzig-Süd 70)	v 01.08.77
7253	PKP *Ty2-806*	
7254		
7255	DB	+ 18.10.54
7256	USSR (Kln 58, Lw 60), (64) JZ 33-295	
7257	PKP *Ty2-807*	*+ 06.06.84*
7258	PKP *Ty2-808*	*+ 14.06.89*
7259	PKP Ty2-809 (Nysa 87)	
52 7260	DB	+ 18.10.54
7261	PKP *Ty2-810*	*+ 29.11.90*
7262	PKP *Ty2-811*	*+ 26.06.78*
7263	DB	+ 18.10.54
7264	DB	+ 18.10.54
7265	DB	+ 18.10.54
7266	PKP *Ty2-812*	*+ 03.05.86*
7267	DR	+ 28.02.51
7268	PKP *Ty2-813*	*+ 13.09.88*
7269	CSD	
52 7270	JDZ 33-152	
7271	PKP *Ty2-814*	*+ 11.04.84*
7272	USSR (Okt 53,Li 59),(63) PKP Ty2-1267(Korsze92)	+92
7273	USSR (Kln 51, Okt 61)	i 73
7274	PKP	

52 7275 USSR (SW 51, i 56 Kohlenindustrie)
7276 USSR (BL 47, W 52, Kln 53, Be 57) i 62
7277 CCCPm → PKP? *Ty2-1196*
7278 ÖStB, ÖBB + 01.12.47
7279 PKP *Ty2-816* *+ 01.06.87*
52 7280 CCCPö
7281 JDZ 33-153
7282 DR + 03.07.70
7283 SNCF 150 Y 7283
7284 DR 52 8153 (Hoyerswerda 70) + 14.02.74
7285 TCDD 56536 (Balikesir 75, Halkapinar 88)
7286 TCDD 56537 (Afyon 87)
7287 TCDD 56538 (Afyon 75, Halkapinar 87)
7288 TCDD 56539 (Erzurum 88)
7289 TCDD 56540 (Afyon 75, Kars 87)
52 7290 TCDD 56541 (Afyon 75, Sivas 87)
7291 TCDD 56542 (Erzurum 75)
7292 TCDD 56543 (Erzurum 75, Erzurum 87)
7293 USSR, (62) PKP Ty2-1216 (Tomaszow 91)
7294 JDZ 33-154
7295 ÖStB, ÖBB (Knittelfeld 63) + 10.02.72 § 985.221
7296 USSR (Ki 48), (64) BDZ 15.212 (Asenowo 92)
7297 CSD 555.081, 555.3081 + 71
7298 FS, (47) DB + 21.04.52
7299 SWDE DB (Friedrichshafen 50) + 18.10.54
52 7300 JDZ 33-207
7301 DB + 18.10.54
7302 USSR (Li 45, BL 47, Li 59), (63) PKP Ty2-1300 + 85
7303 DB + 18.10.54
7304 DR (Neustrelitz 70) + 03.07.79
7305 USSR 1042.537 (Kaliningrad 85) + 90
7306 USSR 1042.538 (Radviliskis 85) + 85
7307 PKP Ty2-817 (Chabowka 88) + 07.06.91
7308 PKP *Ty2-818* *+ 22 02.88*
7309 DR (Elsterwerda 70) + 14.11.79
52 7310 USSR (Len 51, La 57, i 58 Kohlenindustrie)
7311 USSR (Kln 51, Lw 59), (63) MAV 520,013 + 03.71
7312 USSR (Li 47, Mo 61), (64) BDZ 15.254 (Kalo'wetz 92)
7313 DR 52 8138 (Stendal 70) z 11.87 M
7314 USSR (Od 50, Li 59), (63) PKP
7315 CCCPc (CSD 555.0151 ?)
7316 USSR (Len 51, Lw 61), (64) JZ 33-282
7317 USSR, Be 1042.539 (Berestowiza 93)
7318 USSR (Be 50, Li 58) i 58
7319 USSR (BL 50, Kln 54) + 01.05.71
52 7320 USSR, Uk
7321 USSR (Li 49, La 56) + 01.11.74
7322 USSR + 85
7323 USSR (Li 49), (62) DR (Hoyerswerda 70) v 14.08.76
7324 USSR 1042.542 (Vilnius 85) + 88
7325 USSR, Uk 1042.544 (Wjerbka 89)
7326 PKP *Ty2-819* *+ 03.11.89*
7327 PKP Ty2-957 (Korsze 92) + 17.08.90
7328 PKP *Ty2-821*
7329 DR (Berlin Lichtenberg 70) + 30.01.80
52 7330 JDZ 33-155
7331 USSR (Kow 45), (63) MAV 520,045 + 08.69
7332 PKP Ty2-822 (Zakopane 91) + 15.12.92
7333 PKP *Ty2-823*
7334 PKP Ty2-824 (Chelm 92)
7335 USSR (Kow 50), (63) MAV 520,007 # + 05.79
7336 DR + 20.12.51
7337 PKP *Ty2-825* *+ 21.02.88*
7338 USSR (BL 48, Mo 59), (64) BDZ 15.194 (Kalo'wetz 92)
7339 DB + 18.10.54
52 7340 PKP *Ty2-826* *+ 30.12.89*
7341 USSR 1042.545 (Tschernjachowsk 85)
7342 USSR (Li 47, Okt 54) i 79
7343 USSR (Kow 45, i 67 Kupjansk)
7344 USSR (BL 50, Kln 54) + 01.04.77
7345 USSR (Ungeny 94)
7346 USSR 1042.547 (Krasnowka 93)
7347 USSR, Uk 1042.548 (Rawa Russkaja 92)
7348 USSR (Od 50, Li 58), (63) PKP Ty2-1374 + 08.01.86
7349 USSR (Od 48, Li 59), (63) PKP
52 7350 USSR (Lw 46)
7351 USSR, Be 1042.549 (Lida 93) M
7352 USSR 1042.550 (Krasnowka 93)
7353 USSR (Lw 53), (64) BDZ 15.195 (Asenowo 92)
7354 USSR (Kow 46, Od 53, Ba 54, Es 56) i 64
7355 USSR (Len 51) i 56
7356 USSR, Uk 1042.551 (Grodowiza 92)
7357 USSR (Kln 51, Mo 60), (63) MAV 520,036 + 11.72
7358 ÖStB, ÖBB (St.Pölten 63) + 12.11.75
7359 JDZ 33-208
52 7360 USSR (Lw 46), (63) CSSR
7361 USSR (W 51, Be 57, Lw 59) + 01.07.82
7362 USSR, (62) CSD 555.0223 555.3223 + 68
7363 USSR (Mo 60), (64) BDZ 15.198 (Asenowo 90)
7364 CCCPm (CFR 150.1122)?
7365 USSR (Len 51, Li 58), (63) PKP Ty2-1263 + 92
7366 ÖStB, ÖBB (Knittelfeld 63) + 28.09.73
7367 USSR 1042.552 (Kaliningrad 93)
7368 USSR (Lw 53, Mo 59), (64) BDZ 15.199
7369 CCCPö
52 7370 CCCPc
7371 USSR, (62) CSD 555.0224 555.3224 + 69
7372 USSR (Len 51, i 57) + 01.12.72
7373 USSR, Be 1042.553 (Brest-Wostok 93)
7374 USSR (Ki 44, Lw 51) + 51
7375 USSR (Be 58), (63) PKP Ty2-1338 (Jarocin 89) + 91
7376 USSR 1042.554 (Tschernjachowsk 85) + 85
7377 USSR (Len 51) i 57
7378 FS, (48) DB + 21.04.52
7379 USSR (W 45, Od 49, Mo 58), Export
52 7380 ÖStB, ÖBB (Wels 63) + 17.02.75
7381 DB + 18.10.54
7382 USSR (Kln 51, Be 60), (63) PKP Ty2-1241
7383 ÖStB, ÖBB (Mürzzuschlag 63) + 12.11.75
7384 DB + 18.10.54
7385 PKP *Ty2-827* *+ 25.05.73*
7386 USSR 1042.555 (Zheleznodorozhnyj 93)
7387 USSR (Kow 46, BL 48, Lw 53) + 01.04.79
7388 CSD, 555.0152, (59) BDZ 15.109
7389 USSR (Lw 46, Be 47, Li 57) + 01.12.72
52 7390 USSR (Od 50, Ba 63) + 01.02.77
7391 DB + 18.10.54
7392 USSR (Len 51, i 57 Kohlenindustrie)
7393 ÖStB, ÖBB (Knittelfeld 63) + 16.08.76
7394 PKP *Ty2-828* *+ 01.12.85*
7395 USSR (Ungeny 94)
7396 USSR (Lw 44), PKP *Ty2-829* *+ 07.06.91*
7397 USSR (SW 50, Es 57) i 66
7398 PKP *Ty2-830* *+ 10.04.87*
7399 USSR 1042.557 (Zheleznodorozhnyj 93)
52 7400 PKP *Ty2-831* *+ 02.12.90*
7401 CCCPd
7402 USSR (BL 50, Kln 54, Mo 61) + 01.08.76
7403 FS, (48) DB + 18.10.54
7404 USSR 1042.558 (Mamonowo 92)
7405 ÖStB, ÖBB (St.Valentin 63) + 29.12.75
7406 USSR (Len 51, Be 62), → Pi (Szczakowa 91)
7407 USSR, (63) CSD 555.0265 555.3265 + 68
7408 ÖStB, ÖBB (Knittelfeld 63) + 22.11.72
7409 ÖStB, ÖBB (St.Veit/Glan 63) + 10.78 M
52 7410 USSR 1042.559 (Krasnowka 93) + 89
7411 USSR, Uk 1042.560 (Rawa Russkaja 92)
7412 ÖStB, ÖBB (Straßhof 63) + 15.06.73
7413 USSR (Okt 58, Be 60), (63) PKP Ty2-1255
7414 USSR, (63) CSD 555.0284 555.3284 + 68
7415 ÖStB + 11.01.47
7416 USSR 1042.561 (Vilnius 85) + 88
7417 USSR, Be1042.562 (Berestowiza 93)
7418 USSR (Lw 46, La 47) + 01.07.71
7419 USSR (Kow 50, i 56 Kohlenindustrie)
52 7420 DR(w) KrV + 01.10.46
7421 DB + 21.04.52
7422 USSR (Lw 46, Li 47) + 01.11.73
7423 ÖStB, ÖBB (St.Pölten 63) + 20.10.72
7424 USSR (Be 50, SW 59, Lw 61), (64) JZ 33-283

52 7425 TCDD 56544 (Afyon 75, Sivas 88)
7426 TCDD 56545 (Erzurum 75, Erzurum 90)
7427 TCDD 56546 (Erzurum 75, Erzurum 88)
7428 TCDD 56547 (Izmir 75) M
7429 TCDD 56548 (Balikesir 75, Halkapinar 95)
52 7430 TCDD 56549 (Afyon 75, Izmir Alsancak 95)
7431 TCDD 56550 (Afyon 75, Afyon 90)
7432 TCDD 56551
7433 TCDD 56552 (Balikesir 75, Afyon 95)
7434 TCDD 56553 (Balikesir 75, Balikesir 84) M

---

52 7435 - 52 7527

ŠKODA

52 7435 - 7439 11/43 - 12/43
52 7440 1/44
52 7441 12/43
52 7442 - 7524 1/44 - 12/44
52 7525 - 7527 2/45

---

52 7435 PKP *Ty2-832* *+ 16.06.86*
7436 USSR 1042.563 (Radviliskis 85) + 85
7437 CSD 555.013 + 72
7438 USSR (Ungeny 94)
7439 DB + 10.08.57
52 7440 CSD 555.014 555.3014 + 69
7441 USSR 1042.565 (Tallinn 85) + 87
7442 DB + 21.04.52
7443 USSR, (63) MAV 520,018 (64) GySEV + 01.10.79
7444 SNCF 150 Y 7444
7445 FS, (48) DB + 10.08.57
7446 DB + 10.08.57
7447 CSD 555.008 555.3008 + 72 M
7448 DB + 10.08.57
7449 DB + 10.08.57
52 7450 CSD 555.015 555.3015 + 72
7451 DB + 25.10.54
7452 CSD 555.016 + 70
7453 CSD 555.017 + 70
7454 CSD 555.018 555.3018 + 70
7455 USSR (Li 46, Be 46, Kow 46, Kln 51, Li 58),
(62) DR (Dresden 70) + 20.02.75
7456 CSD 555.019 + 71
7457 CSD 555.020 555.3020 + 69
7458 CSD 555.021 555.3021 + 68
7459 CSD 555.022 + 70
52 7460 CSD 555.023 555.3023 + 68
7461 CSD 555.024 + 67
7462 USSR (Li 47) + 01.05.75
7463 DR (Kamenz 70) + 14.04.76
7464 DR (Pasewalk 70) + 03.08.73
7465 CSD 555.025 555.3025 + 72
7466 CSD 555.026 555.3026 + 70
7467 CSD 555.027 555.3027 + 67
7468 CFR 150.1123 (Dej 95) M
7469 CSD 555.028 555.3028 + 70
52 7470 CSD 555.029 555.3029 + 70
7471 CSD 555.030 + 70
7472 CSD 555.031 + 70
7473 ÖStB, ÖBB + 27.11.57, → Ungarn
7474 CSD 555.032 555.3032 + 70
7475 DB (Soltau 50) + 16.12.59
7476 CSD 555.033 555.3033 + 68
7477 USSR (Lw 46, Kow 46, Li 58, i 58 Kohlenind.)
Nach einer anderen Quelle:
52 7477 → CSD 555.034 555.3034 + 69
7478 CSD 555.035 + 71
7479 USSR (BL46, Be 48, Be 58), (63) PKP Ty2-1352 + 89
52 7480 USSR (BL 47), (63) PKP Ty2-1325 (Stalowa Wola 93)
7481 ÖStB, ÖBB + 17.05.57, → Ungarn
7482 ÖStB, ÖBB (Wels 63) + 21.03.75
7483 CSD 555.036 + 71
7484 CSD 555.037, (59) BDZ 15.111
7485 ÖStB, ÖBB + 23.12.57, → Ungarn
7486 ÖStB, ÖBB + 27.12.62. Rahmen → 52.284
7487 CSD 555.038, (59) BDZ 15.107
7488 CSD 555.039 555.3039 + 69
7489 CSD 555.040 555.3040 + 70
52 7490 CSD 555.041 555.3041 + 70
7491 CSD 555.058 555.3058 + 68
7492 DB + 15.11.57
7493 CSD 555.042 555.3042 + 70
7494 CSD 555.043, (59) → BDZ
7495 USSR (Ki 51), (64) BDZ 15.231 (Kalojanowetz 92)
7496 USSR Uk 1042.566 (Rawa Russkaja 92)
7497 USSR (Od 50, Li 58), (63) PKP
7498 CSD 555.044 555.3044 + 70
7499 USSR, (62) CSD 555.0242 555.3242 + 71
52 7500 USSR (Okt 53, Li 58), → PKP Ty2-1275 + 04.11.89
7501 CSD 555.045 555.3045 + 70
7502 ÖStB +46 (CSD 555.046 555.3046) ?
7503 USSR (Lw 52) i 69
7504 USSR (Lw 52, La 58) + 70
7505 USSR (Be 53, Mo 61), (63) MAV 520,025 + 11.75
7506 CSD 555.047 555.3047 + 70
7507 CSD 555.048, (59) → BDZ
7508 PKP Ty2-1140 (Tczew 93)
7509 USSR Uk 1042.567 (Chyrow 89)
52 7510 USSR 1042.568 (Tschernjachowsk 85) + 88
7511 USSR (Be 47, MK 54, Kln 58) i 59
7512 USSR, Be 1042.569 (Wysokoje 93)
7513 USSR (Luninez 90)
7514 USSR (Lw 46), (64) BDZ 15.223 (Bodrowo 92)
7515 CSD 555.049 555.3049 + 68
7516 CSD 555.050 555.3050 + 68
7517 USSR 1042.571 → LG (Subacius 93)
7518 CSD 555.051 555.3051 + 71
7519 CSD 555.052, (59) BDZ 15.119 (Kalojanowetz 92)
52 7520 CSD 555.053 + 64
7521 CSD 555.054 555.3054 + 72
7522 CSD 555.055 + 67
7523 USSR # (→ BDZ 15 272) ?
7524 DB + 10.08.57
7525 USSR (La 47, Ba 53) + 01.09.82
7526 ÖStB, (46) DR(w), DB + 20.07.59
7527 CSD 555.056, (59) BDZ 15.120 (Pernik 90)

---

52 7535 - 52 7539

SCHICHAU

5/44

---

52 7535 PKP *Ty2-1141* *+ 17.03.90*
7536 PKP *Ty2-1193*
7537 USSR (Okt 51, Be 59), (63) PKP Ty2-1257 + 75
7538 PKP *Ty2-1187*
7539 USSR (Lw 53), (64) BDZ 15.200

52 7540 - 52 7748

Wiener Lokomotivfabrik

1/44 - 4/44

---

52 7540	USSR Uk (Wjerbka 89)	
7541	DB	+ 21.04.52
7542	DB	+ 18.10.54
7543	USSR Uk 1042.573 (Rawa Russkaja 92)	
7544		
7545	USSR Uk (Rawa Russkaja 92)	
7546	USSR Be 1042.575 (Brusgi 93)	
7547	PKP *Ty2-835*	*+ 15.05.90*
7548	USSR (La 50, Kln 51, Okt 55)	+ 01.11.76
7549	PKP *Ty2-836*	
52 7550	DR 52 8102 (Frankfurt/Oder 70)	z 02.88 M
7551	USSR 1042.576 (Tschernjachowsk 85)	+ 88
7552	USSR Be 1042.577 (Brusgi 93)	
7553	USSR (Okt 50, Lw 61), (64) JZ 33-284	
7554	DR (Berlin Schöneweide 70)	+ 30.05.80
7555	USSR 1042.578 (Zheleznodorozhnyj 93)	
7556	PKP *Ty2-837*	
7557	USSR (Li 47, Ba 53, Li 56)	i 58
7558	USSR (Be 53, Lw 59), (63) MAV 520,019	+ 04.76
7559	USSR (La 56, Lw 60), (63) MAV 520,046	+ 06.75
	DR 52 5244 (47) → 52 7559" DR +53	
	Nach einer anderen Quelle 52 7559" → USSR.	
52 7560	PKP *Ty2-1142*	
7561	USSR (BL 45, Ki 47, Wnz 51, SW 53)	i 56
7562	USSR 1042.579 (Tapa 85, Valga 88)	+ 90
7563	USSR, (63) CSD 555.0287 555.3287	+ 68
7564	USSR (Kow 46, Lw 53), (63) MAV 520,072	+ 04.79
7565	USSR, (63) CSD 555.0254 555.3254	+ 72 M
7566	DR (Engelsdorf 70)	+ 02.03.78
7567	USSR (i)	
7568		
7569	DB	+ 18.10.54
52 7570	ÖStB, ÖBB (Knittelfeld 63)	+ 20.01.73
7571	USSR, Uk (Rawa Russkaja 92)	
7572	USSR, (62) CSD 555.0234 555.3234	+ 68
7573	USSR Be 1042.581 (Berestowiza 93)	
7574	ÖStB, ÖBB (Straßhof 63)	+ 21.06.71
7575	ÖStB, ÖBB	+ 30.04.48
7576	ÖStB, ÖBB (Graz 63)	+ 29.11.77
7577	ÖStB, ÖBB	+ 15.10.47
7578	ÖStB, ÖBB (Straßhof 63)	+ 25.09.71
7579	FS, (49) DB	+ 18.10.54
52 7580	JDZ 33-209	
7581	DR 52 8197 (Templin 70)	M
7582	DR	+ 17.01.72
7583	DR 52 8034 (Leipzig West 70)	M
7584	USSR (Be 47, SW 48, Lw 61)	
	(62) DR (Wustermark 70)	+ 30.01.80
7585	DB	+ 18.10.54
7586	DR	+ 08.47
7587	SWDE DB (Offenburg 50)	+ 18.10.54
7588	FS, (49) DB	+ 18.10.54
7589	FS, (47) DB	+ 21.04.52
52 7590	FS, (48) DB	+ 18.10.54
7591	FS, (47) DB	+ 21.04.52
7592	FS, (49) DB	+ 18.10.54
7593	ÖStB, ÖBB (Graz 63)	+ 05.76 M
7594	ÖStB, ÖBB (Wien Ost 63)	+ 11.78 M
7595	ÖStB, ÖBB (St.Veit/Glan 63) + 25.10.77	§ 985.231
7596	ÖStB, ÖBB (Wien Ost 63)	+ 11.77 M
7597	ÖStB, ÖBB (Mürzzuschlag 63)	+ 16.12.74
7598	ÖStB, ÖBB (Straßhof 63)	+ 28.04.77
7599	JDZ 33-156	
52 7600	DB	+ 18.10.54
7601	SNCF 150 Y 7601	+ 55
7602	CSD	
7603	SWDE DB (Singen 50)	+ 18.10.54
7604	SNCF 150 Y 7604	+ 55
52 7605	DR 52 8008 (Eberswalde 70)	z 08.88 M
7606	DB	+ 18.10.54
7607	DB	+ 18.10.54
7608	DB	+ 18.10.54
7609	DB	+ 13.12.50
52 7610	CCCPd	
7611	USSR (SW 50, Be 55)	+ 01.04.68
7612	ÖStB, ÖBB (Wels 63)	+ 01.78 M
7613	DR 52 8162 (Stendal 70)	+ 15.03.73
7614	DB	+ 18.10.54
7615	DB	+ 21.04.52
7616	DR (Brandenburg 70)	+ 06.09.75
7617	DR (Elsterwerda 70)	+ 06.09.75
7618	PKP *Ty2-1143*	*+ 23.02.90*
7619	USSR, (63) CSD 555.0294 555.3294	+ 68
52 7620	CSD 555.0153, (60) → 33-502	M
7621	USSR, (63) CSD 555.0288 555.3288	+ 71
7622	CSD 555.0154 555.3154	+ 69
7623	USSR (Len 51, Okt 53, Es 56, Okt 61)	+ 01.04.72
7624	USSR 1042.582 (Rezekne 91)	+ 90
7625	USSR, (63) CSD 555.0259	+ 73
7626	PKP *Ty2-1144*	
7627	PKP *Ty2-1145*	
7628	USSR 1042.583 (Riga 92)	+ 90
7629	ÖStB, ÖBB (Selzthal 63)	+ 20.10.75
52 7630	PKP	
7631	DR (Hoyerswerda 70)	+ 06.02.78
7632	PKP *Ty2-1147*	*+ 04.11.89*
7633	USSR (Li 45, Ba 53, Es 56)	+ 01.04.79
7634	CSD	
7635	CCCPc	
7636	USSR 1042.584 (Vilnius 85)	
7637	USSR (Len 51, Okt 53)	+ 01.01.74
7638	JDZ 33-210	
7639	PKP *Ty2-1148*	*+ 12.08.81*
52 7640	JDZ 33-157	
7641	USSR Uk 1042.585	
7642	USSR (Kln 51, Mo 60), (64) JZ 33-299	
7643	ÖStB, ÖBB (Straßhof 63)	+ 30.03.73
7644	USSR (La 50, Kln 51, Li 59), (63) PKP	
7645	DB	+ 18.10.54
7646	DB	+ 18.10.54
7647	PKP *Ty2-1149*	*+ 23.02.84*
7648	USSR (W 51, Kln 53, Okt 61, i 73 Rybinsk)	
7649	JDZ 33-158	
52 7650	USSR, (63) CSD 555.0279 555.3279	+ 68
7651	ÖStB, ÖBB (Straßhof 63)	+ 01.08.70
7652	CCCPc	
7653	USSR (Lw 46, Ki 48)	i 51
7654	DB	+ 18.10.54
7655	USSR (Lw 53), (64) BDZ 15.177 (Bodrowo 92)	
7656	USSR (Be 53), (64) BDZ	
7657	USSR 1042.586 (Mamonowo 93)	
7658	USSR (Od 56, Mo 60), (63) MAV 520,037	+ 04.64
7659	PKP Ty2-838 (Przeworsk 89)	+ 05.10.90
52 7660	BDZ 15.63 (Gorna Orjahowiza 90)	
7661	BDZ 15.64 (Bodrowo 92)	
7662	BDZ 15.65 (Bodrowo 92)	
7663	BDZ 15.66	
7664	BDZ 15.67 (Bodrowo 92)	
7665	BDZ 15.68	
7666	BDZ 15.69	
7667	BDZ 15.70 (Bodrowo 92)	
7668	JDZ 33-159	
7669	USSR Uk	
52 7670	BDZ 15.71 (Mezdra 90)	
7671	BDZ 15.72 (Bodrowo 92)	
7672	BDZ 15.73	
7673	JDZ 33-221	
7674	BDZ 15.74 (Bodrowo 92)	
7675	BDZ 15.75 (Bodrowo 92)	
7676	BDZ 15.76 (Bodrowo 92)	
7677	BDZ 15.77 (Bodrowo 92)	
7678	BDZ 15.78 (Bodrowo 90)	
7679	BDZ 15.79 (Pernik 90)	

52 7680	PKP	*Ty2-1150*	+ *04.11.89*
7681	DB		+ 21.04.52
7682	JDZ	33-160	
7683	DB		+ 21.04.52
7684	JDZ	33-161	
7685	JDZ	33-162	
7686	USSR	(SW 50, Lw 58), (63) MAV 520,084	+ 09.77
7687	ÖStB,	ÖBB (Wiener Neustadt 63)	+ 27.02.75
7688	USSR	(OK 53, Kln 55, Li 58), (63) PKP Ty2-1372	+ 82
7689	DB		+ 18.10.54
52 7690	JDZ	33-163	
7691	BDZ	15.80	
7692	BDZ	15.81 (Kalojanowetz 92)	
7693	JDZ	33-164	
7694	BDZ	15.82 (Bodrowo 92)	
7695	USSR	1042.588 (Vilnius 85)	
7696	JDZ	33-165	
7697	JDZ	33-166	
7698	JDZ	33-167	
7699	BDZ	15.83 (Stanke Dimitrow 90)	
52 7700	BDZ	15.84 (Asenowo 92)	
7701	JDZ	33-222	
7702	JDZ	33-168	
7703	JDZ	33-169	
7704	JDZ	33-170	
7705	JDZ	33-171	
7706	BDZ	15.85 (Todor Kableschkow 90)	
7707	USSR	(Len 51, Okt 53)	+ 01.12.72
7708	JDZ	33-172	
7709	DR(w),	10/45 → DR(o) → USSR	
52 7710	JDZ	33-173	
7711	JDZ	33-174	
7712	JDZ	33-175	
7713	DB		+ 18.10.54
7714	ÖStB,	ÖBB (Wien Ost 63)	+ 17.02.75
7715	JDZ	33-176	
7716	USSR	(Li 50, Ba 53, Li 56)	+ 01.06.72
7717	DR	52 8058 (Cottbus 70)	M
7718	JDZ	33-177	
7719	DB		+ 18.10.54
52 7720	ÖStB,	ÖBB (Selzthal 63)	+ 12.11.75
7721	USSR	(MK 53, Kln 58, Mo 60), Export	
7722	PKP	*Ty2-1151*	
7723	USSR	(Od 56, Lw 62), (64) JZ 33-315	
7724	DB		+ 18.10.54
7725	USSR	(BL 45, La 48, SW 48, i 57 Kohlenindustrie)	
7726			
7727	ÖStB,	ÖBB	+ 20.11.54
7728	USSR	Uk	
7729	USSR	1042.620 (Krasnowka 93)	
52 7730	USSR	1042.621 (Radviliskis 85)	+ 87
7731	PKP	*Ty2-1152*	
7732	USSR	(Tapa 85)	+ 89
7733	PKP	*Ty2-1153*	+ *17.04.91*
7734	DR	52 8173 (Brandenburg 70)	M
7735	ÖStB,	(6/45) FS, DB	+ 18.10.54
7736	USSR	(SW 47, Od 49, OK 53, i 56 Kohlenindustrie)	
7737	USSR	(SW 59, Lw 61), (64) JZ 33-285	
7738	PKP	*Ty2-1154*	+ *21.03.90*
7739	PKP	*Ty2-1155*	+ *28.12.89*
52 7740	PKP	*Ty2-1156*	+ *01.08.85*
7741	PKP	*Ty2-1157*	+ *06.03.87*
7742	USSR	(Mo 62), (64) BDZ 15.259	
7743	USSR	(Nowosokolniki 79, Ostaschkow 88)	+ 06.05.88
7744	PKP	*Ty2-1158*	+ *23.02.90*
7745	PKP	*Ty2-1159*	+ *20.09.89*
7746	PKP	Ty2-1160 (Jaworzyna 84)	+ 26.07.89
7747	DR	(Salzwedel 70)	+ 02.06.75
7748	USSR	(Be 47, Li 58)	+ 01.11.76

52 7749 - 52 7792

MBA

9/44 - 11/44

52 7749	DR	(Hoyerswerda 70)	+ 09.12.83
7750	ÖStB	KrV	+ 01.09.46
7751	USSR,	(63) CSD 555.0253 555.3253	+ 73
7752	CSD	555.0125	
7753	ÖStB,	ÖBB	+ 30.04.48
7754	USSR	(Lw 50), (63) MAV 520,073	+ 06.77
7755	USSR	(Lw 49, Li 59)	+ 01.08.73
7756	PKP	*Ty2-1161*	
7757	USSR	Uk 1042.592 (Grodowiza 92)	
7758	DB		+ 18.10.54
7759	PKP		
52 7760	CSD	555.074	+ 69
7761	USSR	(BL 45, Lw 49), (63) MAV 520,047	+ 05.76
7762	CCCPc	(12/45) → CCCPö	
7763	PKP	*Ty2-1163*	+ *15.05.86*
7764	CSD	555.0155 555.3155	+ 69
7765	DR	(Wustermark 70)	+ 03.04.78
7766	PKP	Ty2-1164 (Jaworzyna 83)	+ 17.03.90
7767	PKP	*Ty2-1165*	+ *19.10.90*
7768	USSR	(Dno 81, Ostaschkow 88)	+ 03.06.88
7769	PKP	*Ty2-1166*	
52 7770	DB		+ 18.04.56
7771	USSR	(BL 48, La 57)	
		(62) DR, 52 8073 (Halle-G 70)	+ 15.08.88
7772	USSR	(Lw 50), (63) MAV 520,074	+ 02.89
7773	CSD	555.0129	+ 73
7774	JDZ	33-211	
7775	USSR	Uk 1042.594 (Rawa Russkaja 92)	
7776	PKP	*Ty2-1167*	+ *05.09.91*
7777	USSR	(Ba 54, Es 56), (63) MAV 520,5003	+ 08.71
7778	DR	52 8090 (Leipzig West 70)	z 05.88 M
7779	USSR	1042.595 (Tapa 85)	+ 88
52 7780	PKP	*Ty2-1168*	
7781	USSR,	(63) MAV 520,094, (10.63) GySEV	+ 01.10.79
7782	USSR	1042.596 (Tschernjachowsk 85)	+ 87
7783	USSR	(Ki 51, Be 60), (62) PKP Ty2-1231	+ 09.09.87
7784			
7785	PKP	*Ty2-1169*	
7786	USSR	(Lw 50), (63) MAV 520,055	+ 07.72
7787	USSR	Uk 1042.597 (Kowel 94)	
7788	CCCPc		
7789	USSR	(Li 47, La 56, Be 58)	
		(62) DR, 52 8132 (Görlitz 70)	z 08.90 M
52 7790	CSD		
7791	USSR	(Lw 49, La 58), (63) MAV 520,5006	+ 05.70
7792	DB		+ 18.04.56

52 7793

DR 52 1505 (03.53) → 52 7793,
→ 52 8133 (Brandenburg 70) z 04.88 M

52 7794

USSR TE-8013 (62) → 52 7794 + 17.03.70

52.8000

SDZ 33-010 → ÖStB, ÖBB 52.8000 (St.Veit/Glan 63) + 27.10.78

TE-5314 unter Dampf am 26.4.1991 im normalspurigen Reservedepot Wysokoje Litowsk (Weißrußland). Foto: A. Gogolin.

TE-5314 in steam at the standard gauge reserve at Vysokoye Litovsk (Belarus) on 26th April 1991. Photo A. Gogolin.

TE-8026 der Weißrussischen Eisenbahnverwaltung wurde als breitspurige Reservelok erhalten und gehört nun zum Bestand des geplanten Eisenbahnmuseums. Sie stand am 2.5.96 in gut konserviertem Zustand in einem Reservepark an der Strecke Brest—Wysokoje Litowsk abgestellt. Foto: Norbert Tempel.

Broad gauge TE-8026 stored in good condition near Brest for the planned Belarus Railway Museum. Photo 2nd May 1996, Norbert Tempel.

Außerordentlich seltenes Foto einer ehemaligen Lok der DRB Reihe 50 in der Sowjetunion, nähmlich TE-616 (ex DRB 50 616). Sie war eine der letzten sowjetischen 50er und überlebte bis 1966. Sammlung Jukka Nurminen.

An extremely rare photograph of a former German class 50 locomotive in the Soviet Union: TE-616 (ex DRB 50 616). From the collection of Jukka Nurminen.

Supplementary information

52 042 Sold 1.4.76 by DR for industrial use in the Cottbus area.

52 054 Sold 10.9.76 by DR to BMK at Mühlhausen/Thür.

52 069 Recorded in Soviet archives on the Stalin railway 1.1.44 (location not stated).
Taken to Stanislav works (Ivano Frankovsk) and rebuilt as TE-8036.
In August 1951 the loco was delivered to its first depot, Lvov East.

52 082 Recorded in Soviet archives on the Stalin railway 1.5.44 (location not stated).
Recorded at Saki on the Crimean peninsular 1.7.46. In May 1951 taken away for Stanislav works (Ivano Frankovsk), but subsequent fate unknown.
See also 52 3100. It is worth noting that a considerable number of locos were left behind on the Crimean peninsula. A visitor in February 1945 estimated that "a thousand or more" locomotives had been pitched into a chasm at one location. He described it as an amazing sight.

52 216 Sold 27.1.76 by DR to VEB Landmaschinenbau at Tröbitz.

52 234 Sold 17.12.76 by DR to Western Hospital, Stralsund.

52 409 Sold 23.3.77 to agricultural maintenance works at Demmin. Sold in 1980 to VEB Stärkefabrik at Loitz, where it was used until 9.81.

52 444 Withdrawn 20.2.1979. Sold 23.4.79 by DR to VEB Baustoffkombinat, Herzfelde.

52 454 Sold 20.12.77 by DR to VEB Warnow-Werft, Warnemünde.

52 468 Sold by DR 10.9.76 to BMK at Mühlhausen/Thür.

52 573 Sold by DR 18.11.74 to Tagebau Welzow Süd.

52 600 Sold 29.11.71 by DR to VEK Pumpen- und Verdichterbau, Zwickau.

52 776 Recorded at Chop on the Lvov Railway 1.7.46.
Sold to Hungary in 1963 becoming MAV 520,057.
Withdrawn in 1972, but reinstated under the number 520,005. Withdrawn 1974.

52 1180 Given frames from scrapped 52 7139 in August 1972. Scrapped 27.2.76.

52 1268 Sold by DR to Western Hospital, Stralsund.

52 1436 Sold by DR to VEB Warnow-Werft at Warnemünde.

52 1463 Sold 18.5.74 by DR to abattoir at Magdeburg.

52 1466 Sold 3.4.79 by DR to VEB Wohnungsbaukombinat at Rostock.

52 1521 Sold in 1974 by DR to VEB Frottana at Zittau.

52 1549 Taken over by the CSD after the war.
1947 regauged to 1524mm and given number 555.0601.
1964 regauged to 1435mm and given number 555.0185.
Later rebuilt with oil firing and numbered 555.3185.

52 1602 Sold 6.10.76 to VEB Wohnungsbaukombinat (concrete works) at Erfurt.

52 1636 Sold by DR to VEB Weimar-Werk.

52 1646 Sold 27.6.77 to VEB Wohnungsbaukombinat (concrete works) at Erfurt.

52 1725 Taken over by the CSD after the war.
1947 regauged to 1524mm and given number 555.0602.
1961 regauged to 1435mm and given number 555.0180.
Later converted to oil firing and given number 555.3180.

52 1726 Taken over by the CSD after the war and allocated number 555.087.
1949 regauged to 1524mm and given number 555.0605.
1961 regauged to 1435mm and given number 555.0181.
Later converted to oil firing and given number 555.3181.

52 1888 A photo dated 1972 shows this loco had plate frames, whereas the original 52 1888 had bar frames.

52 2011 Sold 17.12.76 by DR to Western Hospital, Stralsund.

52 2202 Sold 7.7.76 to VEB Oberlausitzer Textilbetriebe, Seifhennersdorf; transferred 1978/79 to Spitzkunnersdorf and used until 1987.

52 2312 Rebuilt to 52 8096 by DR in 1963. Sold 7.92 to ÖGEG Linz.

52 2362 Sold 3.11.75 to Kreisbetrieb für Landtechnik, Luckau.

52 2386 Sold 1.1.75 by DR to VEB Spezialbau for use in Leipzig.

52 2399 First recorded in Soviet archives 10.5.44 on Odessa railway.
1.7.46 recorded at Derbent (near the Caspian sea) on Ordzhonikidze railway.
Transferred to Odessa railway in 1949 and to Moldavian railway in 1959.
To Bulgarian State Railways in 1964.

52 2431 Sold 28.7.75 by DR to industrial user at Seifhennersdorf.

52 2451 Recorded at Nautzken (in former East Prussia) on the Lithuanian railway 1.7.46.
Later allocated to Latvian railway (10.5.48), Kishinev railway (10.48), Odessa–Kishinev railway (1.6.53), Odessa railway (1.6.56), and Lvov railway (7.58).
To Hungarian State Railways in 1963 and numbered 520,086.
Withdrawn 11.72 but reinstated as 520,036. Finally withdrawn 5.84.

52 2472 Sold 17.10.75 by DR to furniture factory at Köthen.

52 2501 Rebuilt to 52 8124 by DR in 1965. Sold to ÖGEG Linz 5.92.

52 2630 Sold 27.7.76 by DR to Dauermilchwerke at Stendal.

52 2661 Sold by DR to Western Hospital, Stralsund.

52 2867 Sold 24.10.73 to VEB Technische Textilien, Meerane.

52 3100 First recorded in Soviet archives on the Stalin railway 10.5.44 (location not stated).
Recorded 1.7.46 at Saki on Crimean peninsula. Taken away May 1951 for Stanislav works (Ivano Frankovsk), but subsequent fate is unknown.

52 3145 Sold 2.7.79 by DR to VE Vereinigte Holzwerke at Briescht. Used until 1987.

52 3169 Rebuilt to 52 8088 by DR in 1963. Suffered accident in March 1974, and sold 11.75 for industrial use at Coswig. Scrapped 10.86.

52 3217 Sold 23.10.75 by DR to VEB Reifenwerke, Fürstenwalde.

52 3260 1.10.74 to industrial user at Großschönau.

52 3295 Sold 1.8.77 by DR to Fleischkombinat at Zittau.

52 3420 Rebuilt by DR to 52 8098 in November 1963.
In September 1985 received frames from 52 8101.

52 3443 Transferred from Soviet Union to Hungarian State railways in 1963, becoming 520,049. Withdrawn 11.72 but reinstated as 520,023. Finally withdrawn 12.79.

52 3767 Taken from Germany to USSR in 1947. Recorded on Lvov railway 1.1.48.
Converted to 1435mm gauge February 1958. To Hungary 1963, becoming 520,069. Withdrawn 11.72 but reinstated as 520,065. Finally withdrawn 4.79.

52 4751 Recorded in Soviet archives on Stalin railway 10.5.44, location not stated.
Recorded 1.7.46 at Pyatikhatki, 125 km west of Dnepropetrovsk. Taken to Stanislav (Ivano Frankovsk) works 1950. After overhaul, carried number TE-8026.

52 4926 1.1.85 sold by DR for industrial use at Werneuchen. Used until 1987.

52 4998 Sold 10.11.78 to VEB Interdruck, Leipzig.

52 5004 Sold 27.7.77 by DR to VEB Weimar-Werk.

52 5156 MAV 520,053. Withdrawn 11.72 but reinstated as 520,034. Withdrawn 3.84.

52 5191 Sold 27.7.76 to VEB Getränkekombinat, Magdeburg.

52 5238 Sold 13.10.69 to a dairy at Grevesmühlen.

52 5260 Rebuilt 12.63 by DR as 52 8101. Scrapped 9.85; frames to 52 8098 (now preserved).

52 5410 Sold 20.6.77 by DR to VEB Energieversorgung Spremburg.

52 5758 Sold 27.7.76 by DR to VEB Lackfabrik at Köthen.

52 5960 Sold 17.12.76 by DR to Western Hospital, Stralsund.

52 6128 MAV 520,071. Withdrawn 11.72 but reinstated as 520,006. Withdrawn 8.74.

52 6356 Sold 1975 by DR to GPG at Berlin-Buchholz.

52 6420 Sold 27.06.77 by DR to VEB Wohnungsbaukombinat (concrete works) at Erfurt.

52 6725 Recorded on the Soviet South-Western railway 1.4.44. To Byelorussian railway in July 1945, to Kovel railway September 1946, to Lithuanian railway April 1958.

52 6930 Sold 14.09.76 by DR to Fleischkombinat at Görlitz.

52 6963 Recorded on the Soviet Byelorussian railway in November 1950.
To South-Western railway 21.7.59, then to Lvov railway in March 1961 and to DR 10.2.62. Withdrawn 30.7.75. Sold 23.8.75 to China-clay works at Kemmlitz.

52 7057 Observed derelict at Rava Russkaya in 1992. Broad-gauge loco. Carried incorrect computer-number (1042 515 7).

52 7139 + 27.12.71. Scrapped 26.08.72 at RAW Stendal. Frames to 52 1180.

52 7252 Sold by DR to VEB Energieversorgung at Spremberg.

52 7323 Sold by DR in 1976 to a concrete works at Zeischa.

52 7335 Transferred from Soviet Union to Hungarian State Railways in 1963, becoming 520,007. Withdrawn 11.72 but reinstated as 520,028. Finally withdrawn 5.79.

52 7523 Recorded on 1-7-46 at Korolevo, Lvov Railway.
According to Soviet order TsT-14 dated 26-4-1963 this loco was nominated for sale to Czechoslovakia, but there is no subsequent record of it on the Czech state railways. The Bulgarian author Dimiter Dejanow suggests it became BDZ 15.272 in 1964, though on the other hand it does not appear on the Soviet list of locos transferred to Bulgaria in 1964.

Appendix A: Works numbers

Borsig (Borsig Lokomotiv-Werke, Hennigsdorf bei Berlin)

Borsig works Nr	*Locomotive Numbers*	*Delivery*	*Built by (Factory / Works Nr)*
15446	52 001	09/42	Borsig
15447 – 15620	52 350 – 52 523	04/43-09/43	Borsig [1]
15621 – 15622	52 524 – 52 525	01/44	Borsig
15623 – 15701	52 526 – 52 604	12/43-05/44	Schwartzkopff 13093–13171 [2]
15702 – 15788	52 605 – 52 691	10/43-05/44	Schichau 3964 - [3]
15789 – 15833	52 692 – 52 736	03/44	Henschel 27885–27929
15834 – 15841	52 737 – 52 744	04/44	Henschel 27966–27973
15842 – 15851	52 745 – 52 754	05/44-06/44	Henschel 28011–28020
15852 – 15863	52 755 – 52 766	06/44	Henschel 28333–28344
15864 – 15881	52 767 – 52 784	07/44	Henschel 28052–28069
15882 – 15914	52 785 – 52 817	07/44-08/44	Henschel 28110–28142 [4]
16277 – 16280	SDZ 33.001–33.004	1944	Henschel 28370–28373
16281 – 16291	SDZ 33.005–33.015	1944	Henschel 28041–28051
16292 – 16301	52 4400 –52 4409	08/44	Henschel 28143–28152
16302 – 16312	52 4410 –52 4420	09/44	Henschel 28182–28192

Examples of observed chassis works numbers:

[1] Ty2.116 (52 381)	BLW 15478
[2] Ty2.567 (52 550)	Schwartzkopff 13117
Ty2.853 (52 595)	Borsig 15692
[3] Ty2.858 (52 628)	Borsig 15725 + Schichau 4107
[4] Ty2.882 (52 813)	Henschel 28138

Cegielski / DWM (Deutsche Waffen- und Munitionsfabriken AG), Poznan

Works Numbers	*Delivery*	*Loco Numbers*	*Notes*
514–613	5/43-11/43	52 1100–52 1199	DRB
614–802	11/43-8/44	52 1200–52 1336	DRB
	11/43	52 3632–52 3634	DRB
	11/43-12/43	52 3656–52 3661	DRB
	12/43-2/44	52 3689–52 3694	DRB
	2/44-3/44	52 3723–52 3728	DRB
	3/44-4/44	52 3754–52 3759	DRB
	5/44-6/44	52 3782–52 3787	DRB
	6/44-7/44	52 3819–52 3824	DRB
	7/44-8/44	52 3856–52 3861	DRB
	8/44	52 3880–52 3887	DRB
803–815	8/44-9/44	52 1337–52 1349	DRB
816–881	9/44-1945	52 4500–52 4565	DRB

882	1945	52 4566 ?	*Ty42-31 ?*	→ USSR 52 4566
883	1945	52 4567 ?	*Ty42-32 ?*	→ USSR 52 4567
884	1945	*52 4568 ?*	Ty42-33 ?	PKP
885	1945	*52 4569 ?*	Ty42-34 ?	PKP
886	1945	*52 4570 ?*	*Ty42-35 ?*	
887	1945	*52 4571 ?*	*Ty42-36 ?*	→DR 52 1505"

888–932	1945-1946	Ty42-37–Ty42-81	PKP [4]
933–947	1946	Ty42-102–Ty42-116	PKP
948–962	1946	Ty42-132–Ty42-146	PKP

Cegielski / DWM Notes:

1) Where a number is shown in italics, it is not known if a locomotive ever carried this number.

2) Works numbers up to 802 are based on the book by Griebl and Wenzel. Higher works numbers are based on observed work-plates and locomotive and boiler documentation.

3) Between 783 and 805 DWM is thought to have built 24 locomotives although there are only 23 works numbers! See also Franz Kraus, "Bemerkungen zu den Lokomotiven DRB 52 1340 und 2 x 52 1325 von DWM", Eisenbahn-Modelleisenbahn (Wien) 8/93 p 148.

4) Some of these locos originally carried 52 xxxx numbers. For example, the log-book of boiler 2797 of 1945 states that the boiler was originally on locomotive 52 4574 (Cegielski 890), with the loco number later changed to Ty42-39.

Esslingen

4536–4541	1/43-2/43	52 1350–52 1355
4542	2/43	52 1357
4543	2/43	52 1356
4544–4657	2/43-8/43	52 1358–52 1471
4696–4783	8/43-2/44	52 1472–52 1559
4784	2/44	52 1562
4785–4786	2/44	52 1560–52 1561
4787–4802	2/44-4/44	52 1563–52 1578
4803–4813	3/44-5/44	52 1585–52 1595
4814	4/44	52 3834
4815	5/44	52 3835

4818–4820	12/43	52 3668–52 3670
4821–4824	12/43-1/44	52 3701–52 3704
4825–4828	1/44-2/44	52 3732–52 3735
4829–4830	2/44	52 3760–52 3761
4831–4834	4/44	52 3797–52 3800
4835	5/44	52 3836
4837	6/44	52 3837
4838–4841	6/44	52 3871–52 3874
4844–4845	1944	52 3762–52 3763 ex 52 1523, 52 1525 (Esslingen 4747, 4749)
4846–4849	5/44	52 1596–52 1599
4850–4855	5/44	52 1579–52 1584
4856–4857	5/44	52 1098–52 1099
4858–4860	5/44-6/44	52 871–52 874

Graffenstaden

7867–8005	6/43-10/44	52 1600–52 1738
8006–8022	1946	SNCF 150 Y 1739–150 Y 1755, (8006, 8009-8011, 8007/8, 8012-8022)
8044–8053	1946	CFL 56.11–56.20

Henschel

26841–26845	9/42-10/42	52 002–52 006
26846–26915	1/43-2/43	52 2090–52 2159
26925–26993	10/42-12/42	52 007–52 075
26994–26999	12/42	52 344–52 349
27011–27100	2/43-4/43	52 2160–52 2249
27178–27355	2/43-7/47	52 1850–52 2027
27356		Class 52 boiler
27418–27734	4/43-10/43	52 2250–52 2566
27735–27744	1943	TCDD 56501–56510
27745–27767	10/43-11/43	52 2577–52 2599
27768–27777	11/43-12/43	52 2567–52 2576
27778–27884	12/43-2/44	52 2600–52 2706
27885–27929	3/44	52 692–52 736
27930–27936	4/44	52 2707–52 2713
27937–27945	1944	HDZ 30-016 – 30-024
27946–27965	4/44	52 2714–52 2733
27966–27973	4/44	52 737–52 744
27974–27999	5/44	52 2734–52 2759
28002–28010	5/44	52 2760–52 2768
28011–28020	5/44-6/44	52 745–52 754
28040	7/44	52 2806
28041–28051	1944	SDZ 33-005–33-015
28052–28069	7/44	52 767–52 784
28070–28109		CFR 150 1061–150 1100
28110–28142	7/44-8/44	52 785–52 817
28143–28152	8/44	52 4400–52 4409
28153–28181	8/44-9/44	52 2807–52 2835
28182–28192	9/44	52 4410–52 4420
28193–28250	9/44-7/46	52 2836–52 2893
28251–28273	1/45-1/47	52 1796–52 1818
28274		Class 52 boiler
28275–28276	1946	30 & 31 Rheinische Braunkohlenwerke
28277–28296	1/48-11/49	52 124–52 143
28297–28312	1949-1951	52 875–52 890
28313–28314	2/51	42 9000–42 9001
28315–28316	1951	52 891–52 892
28321–28332	6/44	52 2769–52 2780
28333–28344	6/44	52 755–52 766
28345–28369	6/44-7/44	52 2781–52 2805
28370–28373	1944	SDZ 33-001 – 33-004

Jung

11091–11100	1/43-2/43	52 076–52 085
11111–11331	2/43-12/44	52 3100–52 3320
11332–11342	12/45-11/46	52 3321–52 3331

Krauss-Maffei

16397–16434	2/43-3/43	52 086–52 123
16476–16670	3/43-8/43	52 3350–52 3544
16671–16674	9/43-2/45	52 3620–52 3623
16682–16755	8/43-11/43	52 3545–52 3618
16757	11/43	52 3619

52 3625–52 4049 were originally ordered from Krauss-Maffei. See list of class 52 locomotives.

Examples of observed chassis works numbers:
PKP Ty2-1025 (52 3870) = 17065
MAV 520.020 (52 3897) = 17093

Krenau (Oberschlesische Lokomotivwerke AG) / Chrzanów

1134–1261	4/43-10/43	52 5125–52 5252
1262–1263	10/43-11/43	52 3640–52 3641
1264–1283	11/43	52 5253–52 5272
1284–1287	11/43	52 3671–52 3674
1288–1301	12/43	52 5273–52 5286
1302–1304	12/43	52 3705–52 3707
1305–1320	1/44	52 5287–52 5302
1321	1/44	52 3708
1322–1323	1/44	52 3736–52 3737
1324–1338	2/44	52 5303–52 5317
1339–1340	2/44	52 3738–52 3739
1341	2/44	52 3764
1342–1355	3/44	52 5318–52 5331
1356–1358	3/44	52 3765–52 3767
1359–1369	3/44-5/44	52 5332–52 5342
1370–1372	4/44-5/44	52 3801–52 3803
1373–1392	4/44-5/44	52 5343–52 5362
1393	5/44	52 3804

1394–1395	5/44	52 3838–52 3839	
1396–1407	6/44	52 5363–52 5374	
1408	6/44	52 3841	
1409–1412	6/44	52 3875–52 3878	
1413–1418	6/44-7/44	52 3901–52 3906	
1419–1425	7/44	52 3919–52 3925	
1426	7/44	52 3950	
1427–1437	7/44-8/44	52 818–52 828	
1438	8/44	52 3840	
1439–1443	8/44	52 3951–52 3955	
1444–1457	8/44-9/44	52 829–52 842	
1458–1463	9/44	52 3956–52 3961	
1464–1491	9/44-11/44	52 843–52 870	
1492–1505	12/44-1/45	52 4360–52 4373	
1506–1533	1945	Ty42.1–Ty42.28	*
1559–1566	1945	Ty42.82–Ty42.89	*
1567–1578	1946	Ty42.90–Ty42.101	*
1601–1615	1946	Ty42.117–Ty42.131	*
1642?–1645	1946	Ty42.147–Ty42.150	*

* Note: All details about Ty42 locos built by Chrzanów should be regarded as subject to confirmation.

MBA: Maschinenbau und Bahnbedarf AG, Berlin. (Formerly Orenstein und Koppel)

13801–13868	1/43-4/43	52 4750–52 4817
13869–13871	4/43-5/43	52 4821–52 4823
13872–13880	4/43-5/43	52 4818, 4824, 4819, 4825, 4826, 4820, 4827, 4828, 4837
13881–13883	5/43	52 4838, 4829, 4830
13884–13893	5/43-6/43	CFR 150.1031–150.1040
13894–13895	5/43	52 4839–52 4840
13896–13898	5/43	52 4831–52 4833
13899–13907	6/43	52 4841, 4842, 4843, 4844, 4845, 4846, 4847, 4848, 4834
13908–13910	6/43	CFR 150.1041–150.1043
13911–13920	6/43	52 4849, 4850, 4835, 4851, 4852, 4853, 4854, 4855, 4856, 4836
13921–13934	6/43-7/43	52 4857–52 4870
13941–14036	8/43-1/44	52 4871–52 4966
14037–14045	1/44-2/44	52 4976–52 4984
14046–14054	2/44	52 3635, 3636, 3637, 3662, 3663, 3664, 3665, 3666, 3667
14055–14071	2/44-3/44	52 4985–52 5001
14072–14080	3/44	52 3695, 3696, 3697, 3698, 3699, 3700, 3729, 3730, 3731
14081–14096	3/44-4/44	52 5002–52 5017
14097–14102	4/44	52 3788–52 3793
14103–14116	4/44	52 5018–52 5031
14117–14119	5/44	52 3794–52 3796
14120–14128	5/44	52 3825–52 3833
14129–14135	5/44	52 3862, 5058, 5059, 5060, 5061, 5062, 3863
14136–14140	5/44	52 5063, 5064, 5065, 5066, 3864
14141–14144	5/44-6/44	52 5067–52 5070
14145–14149	6/44	52 3865, 3866, 3867, 3868, 5071
14150–14159	6/44	52 3869, 5072, 3870, 3892, 5073, 5074, 3893, 5075, 3894, 3895
14160–14169	6/44	52 5076, 3896, 3897, 3898, 3899, 3900, 5077, 3913, 3914, 3915
14170–14179	6/44-7/44	52 3916, 5078, 3917, 5032, 5079, 3918, 5033, 5034, 5035, 5036
14180–14184	7/44	52 5037, 5038, 5039, 5040, 5041
14185–14187	7/44	52 5042 ?, 5043 ?, 5080 ?
14188–14194	7/44	52 5044, 5081, 5045, 5046, 5047, 5048, 5049
14195	7/44	52 5082 ?
14284–14286	7/44-8/44	52 5050, 5051 ?, 5083 ?
14287–14289	8/44	52 5052, 5053?, 52 5054
14290–14292	8/44	52 5084, 5055 ?, 5057
14293–14295	8/44	52 5056?, 5085?, 5098?
14296–14305	8/44	52 3938, 3939, 3940, 3941, 3942, 3943, 5099, 5100, 5101, 5102
14306–14315	8/44	52 5086, 5103, 5104, 5105, 5106, 5107, 5108, 5109, 5110, 5111
14316–14324	9/44	52 5112, 5113, 5114, 5087, 5115, 5116, 5117, 5118, 5119
14325–14335	9/44	52 508X, 509X, 512X, 77XX
14336	9/44	52 7753
14337–14339	9/44	52 509X, 51XX, 775X
14340–14358	9/44-10/44	52 7756–52 7774
14359–14360	10/44	52 5091, 7775
14361–14366	10/44	52 7776–52 7781
14367–14371	10/44	52 778X, 509X
14372–14378	10/44-11/44	52 7785–52 7791
14379–14380	11/44	52 5094–52 5095
14381–14382	11/44	52 7792 ?, 5096 ?
14383	11/44	52 5097

Examples of observed chassis works numbers:
14188 = Ty2-1392 (52 5044), 14284 = Ty2-1063 (52 5050), 14329 = Ty2-1086 (52 5123), 14383 = Ty2-1076 (52 5097).

Note however that according to the locomotive log-book 52 5050 = MBA 14195.

Schichau

Correct allocation of Schichau works numbers is not known.

3653–4244	1/43-5/44	52 5375–52 5874, 52 605–52 691, 52 7535–52 7539

Schwartzkopff / BMAG

12020–12030	2/43	52 230–52 240
12149–12234	11/42-2/43	52 144–52 229

12301–12390	2/43-4/43	52 5875–52 5964
12391–12400	1943	CFR 150.1021–150.1030
12401–12420	4/43-5/43	52 5965–52 5984
12426–12685	5/43-10/43	52 5985–52 6244
12688–12764	10/43-12/43	52 6245–52 6321
12765–12774	1944	HDZ 30-001–30-010
12775–12817	1/44-2/44	52 6322–52 6364
12918–12987	2/44-5/44	52 6365–52 6434
12988–12992	1944	HDZ 30-011–30-015
13093–13171	12/43-5/44	52 526–52 604

Škoda

1419–1421	5/43	52 6625–52 6627
1437–1443	5/43-6/43	52 6628–52 6634
1444–1460	1943	CFR 150.1044–150.1060
1461–1510	7/43-11/43	52 6635–52 6684
1511–1533	11/43-2/44	52 7435–52 7457
1556–1625	3/44-2/45	52 7458–52 7527

Wiener Lokomotivfabrik AG (Floridsdorf)

9613–9715	11/42-1/43	52 241–52 343
16116–16143	2/43	52 6685–52 6712
16144–16163	1943	CFR 150.1001–150.1020
16164–16215	2/43-3/43	52 6713–52 6764
16218–16887	3/43-1/44	52 6765–52 7434
16888–17096	1/44-4/44	52 7540–52 7748
17218–17224	10/43-11/43	52 3625–52 3631
17225–17238	11/43	52 3642–52 3655
17239–17252	12/43-1/44	52 3675–52 3688
17253–17266	1/44-2/44	52 3709–52 3722
17267–17280	2/44-4/44	52 3740–52 3753
17281–17293	4/44-5/44	52 3769–52 3781
17294–17307	5/44	52 3805–52 3818
17308–17321	5/44	52 3842–52 3855
17322	5/44	52 3879
17323–17325	6/44	52 4045–52 4047 ?
17326–17327	6/44	52 4048–52 4049

Works plate from 52 3846.

Plate-frame chassis built by Borsig. Borsig's early class 52 locomotives had two large cut-outs over each pony-wheel, giving access to the springs of the leading axle. All standard plate-frame locos had these springs above the level of the buffer beam. Photo: P.J.S.

Appendix B. The identities of the Polish Ty2

1. Ty2-1 – Ty2-1206

1		2		3	
Ty2-115	52 377	Ty2-853	52 595	Ty2-76	52 2757
Ty2-116	52 381	Ty2-858	52 628	Ty2-137	52 6273
Ty2-215	52 1508	Ty2-873	52 750	Ty2-180	52 3256
Ty2-218	52 1600	Ty2-878	52 775	Ty2-197	52 7151
Ty2-256	52 2279	Ty2-882	52 813	Ty2-339	52 2641
Ty2-262	52 2311	Ty2-916	52 1705	Ty2-430	52 2263
Ty2-266	52 2336	Ty2-927	52 2710	Ty2-432	52 3813
Ty2-297	52 2506	Ty2-930	52 2730	Ty2-567	52 550
Ty2-410	52 4784	Ty2-939	52 2772	Ty2-609	52 1484
Ty2-620	52 5957	Ty2-953	52 2817	Ty2-709	52 6395
Ty2-632	52 6027	Ty2-954	52 2826	Ty2-815	52 586
Ty2-637	52 6047	Ty2-981	52 3273	Ty2-820	52 7091
Ty2-644	52 6094	Ty2-985	52 3290	Ty2-847	52 462
Ty2-690	52 6293	Ty2-992	52 3313	Ty2-869	52 5958
Ty2-702	52 6340	Ty2-1000	52 3715	Ty2-871	52 428
Ty2-703	52 6403	Ty2-1012	52 3770	Ty2-891	52 6925
Ty2-755	52 6871	Ty2-1025	52 3870	Ty2-921	52 3284
Ty2-761	52 6906	Ty2-1140	52 7508	Ty2-957	52 7327
Ty2-762	52 6907	Ty2-1160	52 7746	Ty2-967	52 6914
Ty2-765	52 6918			Ty2-976	52 2786
Ty2-770	52 6995			Ty2-1035	52 6191
Ty2-771	52 7001			Ty2-1098	52 6143
Ty2-809	52 7259			Ty2-1124	52 698
Ty2-817	52 7307			Ty2-1129	52 6208
Ty2-822	52 7332				
Ty2-824	52 7334				
Ty2-838	52 7659				

This table shows the original German numbers of Ty2 locomotives as derived from their chassis works-numbers, based on observations made 1983-1993. **Column 1** shows locomotives built in 1943 (with some exceptions) which conform with the 1947 PKP renumbering plan as outlined in table 6. Similarly **column 2** lists locos built in 1944. **Column 3** lists locos which do not fit the renumbering plan at all.

Locomotives have been included in this table only if their original 52 number can be deduced with a high degree of certainty. Worksnumbers from DWM, Oberlok and Schichau have not been included.

The proportion of locomotives listed above which do not agree with the 1947 renumbering plan is about 35%.

Significantly, all the class 52 numbers in column 3 appear in the renumbering plan against other Ty2 numbers. This suggests that whatever caused the nonconformance with the plan took place before the arrival of Ty2-1207–Ty2-1406 from the Soviet Union in the winter of 1962-63.

The reasons why some locos had "wrong numbers" can perhaps be found in their locomotive and boiler documentation. At the time of writing, quite a few of these locos still survive, including

Ty2-29 (chassis Schichau 4002, boiler Borsig 15598),
Ty2-130 (Henschel),
Ty2-406 (Henschel),
Ty42-148 (chassis Jung 11331).

2. Ty2-1207–Ty2-1406

	1	2	3	4	5	6	7	8	9	10
Ty2-1207 - 1210	52 -	-	-	-	-	-	6148	-	3798	6861
Ty2-1211 - 1220	1244	6797	-	839	6799	7293	-	7190	5083x	4362x
Ty2-1221 - 1230	7205	1801	3642	5516x	5650x	6083	-	5302	3605	5348
Ty2-1231 - 1240	7783	7042	5291	6029	-	529	1347	-	3753x	-
Ty2-1241 - 1250	7382	-	1456	-	6200	5093x	3185	402 x	3639x	1616
Ty2-1251 - 1260	6937	810	1497	2376	7413	3724	7537x	2821*	4563x	2526
Ty2-1261 - 1270	2417	5048	7365	3919	2518	3170x	7272	5147	6883	6169
Ty2-1271 - 1280	4941	560	3651	-	7500s	-	6119s	3214s	-	-
Ty2-1281 - 1290	4919s	3858s	3275s	-	6058s	-	-	-	1333	456
Ty2-1291 - 1300	3679s	-	5532s	-	-	5507x	549x	3286	850	7302s
Ty2-1301 - 1310	52 -	3402	772	1645	3952	7082*	-	-	-	6949
Ty2-1311 - 1320	8022*	-	2612	2726	2535	7212x	-	-	6812	-
Ty2-1321 - 1330	-	6804	2234	3422x	7480	2635	6162	6054	-	-
Ty2-1331 - 1340	7083	2383	-	7023	-	5070	-	7375	5986	4891
Ty2-1341 - 1350	-	-	-	-	2197	3736	-	6662	4956	-
Ty2-1351 - 1360	3883s	7479x	6410s	-	3759x	6994	5887	3786	624x	6260
Ty2-1361 - 1370	5566x	8018*	2682	3481	5326x	-	1495x	1519	1226	-
Ty2-1371 - 1380	689	7688s	463	7348s	1410	-	2873	-	-	-
Ty2-1381 - 1390	2361	-	6913x	676	-	2411	2277	-	820	-
Ty2-1391 - 1400	2767	5044	1461	510	3213	-	5141	-	-	2515x
Ty2-1401 - 1406	52 1481	-	1511	-	2393	851	-	-	-	-

* Ty2-1258 (ex 52 2821) carried the number TE-2021 in the USSR.
* Ty2-1306 chassis works number WLF 16498, ex 52 7045 (observation dated 1987).
* Ty2-1311: All evidence suggests Ty2-1311 was TE-8022, but this has not been confirmed. See Appendix D.
* Ty2-1362 originally had boiler 12335 (ex 52 5909), according to PKP records dated 1963.

The Soviet/Polish renumberings listed above are in most cases derived from official PKP records. However these are not always entirely clear, and in these cases the 52 number is marked with an 'x'. Locomotives marked with an 's' come from the records of the late Hans Sternhart of Vienna, and are broadly in agreement with other sources of information. Official records in Russia confirm that all the locomotives shown above were actually transferred from the USSR to Poland in the winter of 1962/63.

It should be added that although the Soviet records for TE-2515 show it went to Poland, it is not included in the list of 200 locomotives actually handed over at the border. The story of this loco continues in the archives of the PKP, which show that in February 1963 boiler 27683 (originally on 52 2515) was recorded on the newly renumbered Ty2-1400.

Observations of chassis worksnumbers confirm the identities of the following locos: Ty2-1216, Ty2-1234, Ty2-1258, Ty2-1260, Ty2-1267, Ty2-1269, Ty2-1298, Ty2-1310, Ty2-1314, Ty2-1325, Ty2-1334, Ty2-1336, Ty2-1338, Ty2-1340, Ty2-1368, Ty2-1371, Ty2-1373, Ty2-1381, Ty2-1384, Ty2-1387, Ty2-1392, Ty2-1397.

Two interesting points arise from the observed works numbers of Ty2-1207–Ty2-1406:

1) No locomotives in this series have been found with frames from locos with lower numbers. This suggests that (at least since 1963) very few Ty2s have swapped frames.

2) The evidence also suggests that Soviet locomotives only very rarely swapped either boilers or frames. 52 2515 is possibly an exception to this generalisation.

3. Soviet records.

Listed below are Soviet records of locomotives transferred to Poland at the location and date indicated.

Source: Central State Archive of the National Economy of the USSR, Volume 1884, List 43, File Nr.744.

Brest, 17 - 20 December 1962.

52 839	6773	6148	3798	6677	6861	1244	6797	6799	7293

Berestovitsa, 17 - 21 December 1962.

52 5083	4362	1801	768	7190	7205	5650	5516	3642	6083
5302	3452	5348	3605	7783	7042	5291	6029	2224	529

Berestovitsa, 4 - 11 January 1963.

52 1347	3753	6301	7382	5421	5538	868	1456	6200	5093
6937	1616	3639	402	3185	7413	3724	2376	810	1497
2417	2526	4563	5048	7537	2021				

Kaliningrad, 12 January - 20 February 1963.

52 5147	7272	3170	2518	6883	7365	560	3651	3919	4941
6169	5566	6260	624	3786	5887	2141	8018	3481	2682
5326	463	1410	7348	7688	3866	811	2361	2873	5604
7314	689	1226	1495	1519	3801	820	2411	4884	2277
2283	676	2767	5619	6772	6913	1461	2655	3213	5044
510	7644	1481	5141	795	2393	1511	6289	851	7497

Chernyakhovsk, date unknown.

52 6119	2623	3214	7500	4409	3858	3275	4919	3567	3253
6058	1653	4554	5237	5670	1333	5532	3679	456	1334

Chernyakhovsk, 5 - 9 January 1963.

52 5507	3286	549	4781	2519	850	7302	3402	7349	772
1645	7082	3952							

Brest, date unknown.

52 2589	623	2624	6949	8022	7212	2482	2535	2612	2726
466	6812	8007	6387	6863	3422	2234	7480	6804	6054
3779	2635	6162	774	7023	6954	7083	2383	5000	7375
5070	7021	4891	759	8011	5986	602	7045	2197	3736
6662	2172	4956	1160	7479	3883	6410	5362	6994	3759

4. Other locos exported from the USSR to Poland.

These were subsequently used on Polish industrial lines and retained their German / Soviet numbers, e.g. Ty2-052, Ty2-2118.

52 052	2118	2521	3458	3494	3554	4504	4544	4925	5309
5472	5593	5680	5903	6690	6771	7060	7155	7173	7406

PKP Ty2-104 at Korsze depot in the summer of 1990. This was a former condensing locomotive, as can be seen by the higher than usual running-plate, and the position of the air reservoirs over the second and third driving wheels. Photo: P.J.S.

Appendix C. Locomotives of the SDZ and HDZ

This table lists the "class 52" locomotives owned by the Serbian State Railways (SDZ) and Croatian State Railways (HDZ) in 1944. Post-war owners and numbers are shown in the second and third columns.

SDZ 33-001 – 33-015
Henschel 1944
28370–28373, 28041–28051

33-001	DB		
002	JDZ	33-002	
003	JDZ	33-003	
004	JDZ	33-004	
005	DB	52 4731	+ 57
006	JDZ	33-006	
007	DB	52 6379"	+ 50
008	JDZ	33-008	
009	JDZ	33-009	
33-010	ÖStB	52.8000	+ 78
011	JDZ	33-011	
012	JDZ	33-012	
013	JDZ	33-013	
014	ÖStB	52.765"	§
015	JDZ	33-015	

HDZ 30-001 – 30-015
Schwartzkopff 1943
12765–12774, 12988–12992

30-001	JDZ	33-016
002	JDZ	33-017
003	JDZ	33-018
004	JDZ	33-019
005	JDZ	33-020
006	JDZ	33-021
007	JDZ	33-022
008	JDZ	33-023
009	JDZ	33-024
30-010	JDZ	33-025
011	JDZ	33-026
012	JDZ	33-027
013	ÖStB +	
014	JDZ	33-029
015	JDZ	33-030

HDZ 30-016 – 30-024
Henschel 1944
27937–27945

30-016	JDZ	33-031	
017	JDZ	33-032	M
018	JDZ	33-033	
019	JDZ	33-034	
30-020	JDZ	33-035	
021	JDZ	33-036	
022	JDZ	33-037	M
023	JDZ	33-038	
024	JDZ	33-039	

52.765" was withdrawn 15.11.1953 and rebuilt into Klima snow-plough 985.200 in 1954.

Newly-painted TE-8021 at Luninets on the Byelorussian railway in 1989. Photo A. Gogolin.

Appendix D. Soviet TE-8000 and TE-9100 series locos

(First published in Continental Railway Journal Nr 90, 1992, and subsequently updated.)

Origins Chassis	/ Boiler	No.	Delivery (Date	/ Rly)	Selected depots	Last known details	Notes
52 3153	0020	TE-8001	4.49	Lvov	Lesnaya	+	B
005	0014	8002	4.49	Lvov	Odessa-Tovarnaya	+	P O
52 5990	52 3558	8003	4.49	Lvov	Krasnovka (93)	1042 599	P V
52 613	0022	8004	5.49	Lvov	Radviliskis	+ 82	P V
[7]		8005	5.49	Lvov	Korolevo	64 BDZ 15.180	P
0025	0024	8006	5.49	Lvov	Kherson	i 56	
0019	0027	8007	5.49	Lvov	Novosokolniki	63 PKP	
0027	0025	8008	5.49	Lvov	Lvov West	+ 11.07.65	
52 3468	006	8009	12.49	Lvov	Nikolayev	+ 64	P
		TE-8010	11.50	Brest-Lit		62 CSD, 555.0232, +29.03.68	
		8011	12.49	Brest-Lit		63 PKP	
1.1.45 Kishinev Rly		8012	10.50	Brest-Lit	Dno	i 57	P
1.4.44 Vinnitsa Rly		8013	4.50	SW	Zhlobin	62 DR 52 7794, +17.03.70	
		8014	12.49	Brest-Lit	Kaliningrad (84)	1042.600, + 91	
		8015	11.49	Brest-Lit	Polotsk	+01.09.72	P V
		8016	12.49	Brest-Lit	Dno	60 BDZ 15.157	P
		8017	4.50	Latvian	Liepaja	+ 01.12.81	B
	52 5909	8018	2.50	Kovel	Novosokolniki	63 PKP Ty2-1362	
		8019	12.50	Kovel	Jelgava	+ 01.08.73	B V
		TE-8020	10.50	Kovel		i 56	
52 5909	[6]	8021	5.50	Lvov	Kaliningrad	1042.601, Neman (93)	B V
[5]		8022	5.50	Odessa		63 PKP Ty2-1311?	
		8023	6.50	Lithuanian	Chernyakhovsk	66 JZ 33.328	P
		8024	10.50	Lvov		62 CSD, 555.0211, +25.06.68	
		8025	10.50	Lvov	Bendery	+ 01.08.74	
ex 52 4751		8026	8.50	Lvov	Chernovtsy	91 Museum loco	B
		8027	1.51	Lithuanian	Rezekne (90)	1042.603, + 90	B
		8028	11.50	Lvov	Korolevo	Khyrov (87)	B
		8029	11.50	Lvov	Chop	65 cement industry	
		TE-8030	11.50	Lvov	Skirotava	+ 01.08.76	P O
		8031	11.50	Lvov	Chernovtsy,	64 BDZ, 15.204, Kaloyanovetz (92),	P [4]
		8032	1.51	Lvov	Korolevo	+ 4.88	
		8033	2.51	Lithuanian	Radviliskis	+ 01.12.73	P V
		8034	3.51	Lvov		[2]	
		8035	[1]				
ex 52 069		8036	8.51	Lvov	Lvov East	+ 01.04.75	B
		TE-9101	2.50	Latvian	Ciekurkalns	i 56	
		9102	1.50	Latvian	Ciekurkalns	+ 01.07.77	B O
		9103	10.49	Lvov		64 BDZ	

Notes:

1] No data exists for any loco with number TE-8035.
2] Records suggest that TE-8034 was still on the Lvov Railway in 1991.
3] Tenders of type K2'2'T32 have been observed with numbers 8048 and 8081.
4] Chassis works number 15531
5] Evidence from Poland suggests TE-8022 became Ty2-1311, but this has not been confirmed. Ty2-1311 had chassis of 52 6319 and (in 1963) boiler of 52 2102.
6] TE-8021 boiler works number 4661 (according to loco's documentation). Chassis works number 12335.
7] Soviet sources suggest frames were from 3564, boiler number "005". Bulgarian sources suggest frames were Schw 12448, boiler ex 52 2309.

(B) Bar-frame
(P) Plate frame

(O) Round casing for steam dome.
(V) Square casing for steam dome.

Appendix E.

Locomotives in the Kaliningrad district in 1992 and 1993

031 Z	061 M	087 B	093 Z	106 B	114 Z	153 A	230 Z	266 Z	307 Z
315 A	607 A	622 g	737 b	777 K	815 K	816 M	858 Z	1102 B	1118 k
1151 b	1158 Z	1170 A	1235 m	1330 B	1342 A	1455 Z	1581 b	1618 M	1687 K
1692 C	1697 A	1713 k	2092 B	2096 A	2144 g	2194 M	2207 C	2232 Z	2247 A
2340 A	2398 B	2698 Z	2742 m	2811 M	2822 Z	3173 m	3176 M	3228 B	3265 M
3356 K	3387 A	3397 g	3419 m	3453 m	3643 A	3644 b	3665 M	3720 B	3744 A
3751 K	3765 Z	3868 Z	3869 Z	3923 A	4363 m	4542 A	4564 C	4754 Z	4796 A
4849 B	4882 M	5221 g	5319 m	5425 M	5483 Z	5508 A	5527 B	5534 K	5597 A
5620 M	5690 Z	5705 m	5731 Z	5886 Z	5890 A	5917 K	5933 b	6004 A	6105 k
6151 B	6164 K	6173 K	6190 A	6223 B	6346 A	6683 B	6725 B	6744 g	6798 Z
6868 m	7074 m	7079 Z	7171 m	7250 M	7346 A	7352 A	7367 K	7386 Z	7399 Z
7404 m	7410 A	7555 Z	7657 M	7729 A	8003 A				

A: Krasnovka
B: Bagrationovsk
C: Chernyakhovsk
G: Gusev
K: Kaliningrad
M: Mamonovo
Z: Zheleznodorozhnyj
Lower case denotes locomotives observed in 1992
Upper case denotes locomotives observed in 1993.

Locomotives in Belarus in 1993

074 V	103 V	205 L	322 N	335 V	424 U	575 V	721 N	804 V	837 L
840 V	1110 V	1128 V	1136	1137 V	1194 V	1204 V	1343 E	1409 V	1631 S
1700 V	1936 E	2099 L	2159 E	2260 U	2297 G	2452 V	2497 E	2522 V	2596 V
2615 L	2719 E	2734 V	2812 V	3161 G	3164 U	3201 V	3269 V	3296 V	3712 V
3826 U	3837 S	3915 V	4373 G	4518 E	4521 V	4543 L	4561 E	4565 E	4811 E
5006 V	5027 V	5033 L	5073 L	5150 L	5248 L	5265 O	5277 V	5314 V	5343 V
5358 V	5408 V	5446	5459 V	5540 V	5653 O	5689 V	5693 N	5806 V	5826 V
5874 U	5892 L	5897 V	5945 V	5973 V	5982 V	6003 L	6057 V	6096 E	6115 L
6365 D	6423 V	6432 V	6433 V	6695 E	6714 V	6719 V	6722 E	6732 V	6782 U
6822 V	6834 V	6921 V	6952 V	7043 S	7123 N	7317 E	7351 D	7373 S	7417 E
7512 V	7546 U	7552 U	7573 E	8021 N					

D: Lida
E: Berestovitsa
G: Grodno
L: Luninets
N: Neman
O: Osipovitchi
S: Brest East
U: Bruzgi
V: Vysokoye Litovsk
Also: TE-1114, 6832, 8026 seen in 1996 at Lyschytzi, north of Brest.

Lokomotives in Ukraine, Summer 1991

080 H	146 E	188 K	205 L	250 ?	257 E	260	380	423 E	457 R
495	528	585 E	636 S	802	831 G	848 R	857 U	863 R	867 R
1103 R	1122 E	1139 P	1210 H	1286 R	1294 R	1322 R	1363 K	1457 R	1510
2408 E	2446 H	2476 R	2588 E	2738 R	2768 E	2791 E	2796 R	2806 E	2878
3127 K	3297 E	3382 P	3385 R	3505 R	3539 R	3556	3607 R	3675 E	3693 R
3710	3718 R	3727	3799 E	3956 H	4361 R	4401 E	4405	4508	4516 R
4519 R	4529 R	4578 R	4812	5217 C	5271 E	5360 P	5424 H	5529 R	5572 G
5610 R	5703 R	5717 H	5919 R	5923 E	5964 E	6036 E	6043 R	6044	6060 E
6076 R	6110 R	6117	6137 G	6179 R	6209 U	6222 K	6270	6274 E	6758
6802	6919 R	6971 E	7030 V	7057 R	7097 G	7133 H	7134 E	7157 E	7164 U
7185 W	7320	7325 E	7347 R	7356 G	7411 R	7496 R	7509 H	7540 E	7543 R
7545 R	7571 R	7641 H	7669	7728 P	7757 G	7775 R	7787 Z	8028 H	8034

The locomotive numbers for the Ukraine are largely based on official information.
The locations of each loco are largely based on railway enthusiasts' observations.

C: Khyrov 1994
H: seen at Khyrov 1987 - 1989
P: Porubnoye 1988 / 89
U: Sudovaya Vishnya 1988
Z: Kovel 1994
E: Vyerbka (Kovel) 1989
K: Korolevo 1987
R: Rava Russkaya 1992
V: Lvov East 1989
G: Grodovitsa 1992
L: Luninets (Belarus) 1993
S: Sudovaya Vishnya 1989
W: Lvov West depot 1993

Appendix F. Rumania

1. Locations of locomotives 1993 - 1995

This incomplete list is based on observations by Roland Beier, David Fakes and Brian Garvin.

150.1012	Brasov
1013	Iasi
1015	Ungheni
1018	Bacau
1021	Bacau
1026	Rosiorii de Vede
1028	Iasi
1031	Lugoj
1032	Rosiorii de Vede
1033	Molid
1035	Bacau
1037	Subcetate
1040	Suceava
1043	Halmeu (B)
1044	Rîmnicu Vîlcea
1046	Lugoj
150.1050	Ungheni
1053	Ciceu
1060	Iasi
1062	Sighetul Marmatiei (B)
1065	Tîrgu Mures
1071	Suceava
1072	Rosiorii de Vede
1074	Ciceu
1075	Oravita
1077	Ungheni
1080	Bacau
1081	Bacau
1085	Rosiorii de Vede
1086	Brasov
1088	Oravita
1093	Rosiorii de Vede
1095	Subcetate
1097	Molid
150.1102	Lugoj
1103	Oravita
1105	Sibiu (M)
1106	Rosiorii de Vede
1107	Turnu Severin
1109	Lugoj
1114	Tîrgu Mures
1118	Rosiorii de Vede
1121	Subcetate
1123	Dej (M)

Abbreviations:

B	1520mm gauge
M	preserved

2. Identities of 150.1101 - 1123

The following list has been suggested by Serban Lacriteanu. Observations of chassis works numbers confirm identities shown with *.

150.1101	52 1385
150.1109	52 077
150.1116	50 3100
150.1120	52 2643
150.1123	52 7468*
150.1103	52 4511*
150.1111	52 6850
150.1117	52 2498
150.1121	52 1127*
150.1105	52 196
150.1114	50 517* (boiler ex 50 306)
150.1118	52 3506
150.1122	52 7364

3. Locomotives hired by the CFR from Germany and still in stock on 23.8.1944

52 020	52 2218	52 3646	52 6009	52 6760
52 170	52 2302	52 4885	52 6050	52 6761
52 257	52 2379	52 5148	52 6075	52 6763
52 498	52 2498	52 5170	52 6115	52 7066
52 1616	52 2607	52 5186	52 6131	52 7121
52 1687	52 2633	52 5462	52 6148	
52 1691	52 3384	52 5717	52 6176	
52 2119	52 3453	52 5945	52 6283	
52 2174	52 3488	52 5964	52 6291	
52 2198	52 3528	52 5973	52 6733	

All locomotives were later regauged and worked on Soviet Railways, with the exception of 52 2498 and 52 6009. The former probably became 150.1117, the fate of the latter is not yet known. This list was kindly supplied by Serban Lacriteanu.

Appendix G. Soviet Locos in 1944

Incomplete list showing the earliest known date and location.

Bialystok Railway										
1.10.44	52 254	679	2319	3895	5419	5422	5427	5497	5595	5598
	52 5650	6116								
Brest-Litovsk Railway										
1.10.44	52 036	081	145	764	2373	5908	6033			
Kishinev Railway										
10.5.44	52 213	265	293	321	1485	1861	2248	3153	3480	4885
	52 5060	5718	5943	6220						
1.10.44	52 2168	2183	2221	3415	6348	7374				
1.1.45	52 019	320	1462	1527	1688	1695	2331	3166	3175	3363
	52 3559	4879	5128	6013	6025	6351	aaaa*			
Latvian Railway										
1.1.45	52 505	2316	5505	6330						
Lithuanian Railway										
1.10.44	52 246	534	5533	6324	6326					
1.1.45	52 405	689	757	2170	2224	5159	5611	6029	6386	6391
Lvov Railway										
10.5.44	52 2383	5151								
1.10.44	52 271	463	532	3391	3575	4805	4846	5307	5507	5672
	52 5791	5794	5935	6048	6244	6247	6993	7396	"5/54" *	
Odessa Railway										
3.44	52 3352k	3385	6114							
1.4.44	52 6292									
5.44	52 037	071	303	359k	369	429	482	1126	1133	1620k
	52 1863	1872	1903	1904	1915	2102	2143	2148	2241	2266x
	52 2276	2286	2321	2399	3106	3165	3427	4815k	5384	5443k
	52 5548	5898k	5901	5910	5993	6041	6098	6108	6132	6166
	52 6690	6756	6923							
1.1.45	52 312	1619	1908	3135	3389					
Southern Railway										
1.4.44	52 319	5407								
South-Western Railway										
4.44	52 034	116	117	2186	2189	2191	2197	6030	6138	6725
	52 6734									
10.5.44	52 2181	5920	5990							
1.1.45	"52 02/1" *									
Stalin Railway										
1.1.44	52 069	215	296	2166	2267	2268	2274	5444	6078	6089
1.4.44 - 1.8.44 (Various dates)										
	52 082	207	311	1372	1875	1880	2378	3100	3150	3351
	52 3358	3359	4751	4781	4809	"5090" *	5131	5401	6097	
Vinnitsa Railway										
4.44	52 089	485	549	1383	1933	2193	2252	2303	3148	3168
	52 3172z	5903	5934	5938	6119	6121	6122	6129	6149	6167
	52 6168	6295	6319	6840	bbbb *					
5.44	52 035k	040	110	241	256	501	611	1191	3421	5875
	52 5955	5986	6150	6281z						
1.1.45	52 cccc *									
Western Railway										
1.1.45	52 2223	3102	3181	3671						
	52 5998	dddd *								

Notes:

k: no boiler x: boiler exploded z: wrecked

Railway headquarters:

Southern: Kharkov South-Western: Kiev
Stalin: Dnepropetrovsk Western: Smolensk

** Locomotives without numbers:*

52 aaaa	later became TE-8012.
52 bbbb	later became TE-8013.
52 cccc	possibly became TE-9103.
52 dddd	Not taken into service. Condemned 1951.

** Locomotives with unidentifiable numbers:*

"02/1"	Fate unknown after 5.48.
"5/54"	Fate unknown.

"52 5090" = 52 6090?

Appendix H. Soviet renumbering 1984

		1	2	3	4	5	6	7	8	9	0
1042 001-010	TE	021	023	031	?	061	068	074	080	087	093
011-020		098	103	106	114	146a	153	155	161	188a	205
021-030		209	214	218	219a	229a	230	239	250a	257	260a
031-040		266	298	307	313	315	318	322	335	340	380a
041-050		389	423a	424	435a	456c	457a	479	495a	500a	507
051-060		517a	519	525	528a	575	585a	607	622	636a	640
061-070		643	646	647	655a	656	687	691	721	736a	737
071-080		769a	777	786a	799	802a	804	815	816	831a	837
081-090		840	848a	856	857	858	863	867	1102	1103	1110
091-100		1114a	1118	1122a	1128	1136a	1137	1139a	1148a	1158	1170
1042 101-110	TE	1194	1204	1210	1223	1235	1263a	1283	1286	1294	1322
111-120		1330	1335a	1342	1343	1363a	1370	1390a	1392	1409	1416
121-130		1417	1455	1457	1469	1470	1476	1479a	1510a	1553a	1554
131-140		1581	1614	1618	1631	1687	1691	1692	1697	1700	1701
141-150		1713	1731a	1738a	1796	1936a	1937	2092	2096	2099	2126a
151-160		2128	2130	2140a	2142	2144	2159	2179a	2188	2194	2196
161-170		2207	2218	2232	2247	2251	2254	2260	2271	2275	2288
171-180		2297	2308a	2340	2363a	2381	2394	2396	2398	2408a	2413
181-190		2434	2446	2452	2476	2497	2505a	2522	2563a	2564	2586
191-200		2588a	2590	2596	2607	2615	2639	2668	2695a	2698	2707
1042 201-210	TE	2717	2719	2725	2727a	2734	2738	2742	2768a	2773	2789
211-220		2790a	2791	2796	2804	2806a	2809a	2811	2812	2822	2878a
221-230		2879	3117a	3127a	3136	3161	3162	3164	3173	3176	3201
231-240		3210a	3228	3265	3266	3269	3297a	3356	3368	3371a	3382a
241-250		3384a	3385	3387	3388a	3397	3419	3434	?	3445a	3453
251-260		3463	3505a	3539	3545	3556a	3607	3643	3644	3665	3667
261-270		3675a	3693	3708a	3710a	3712	3718	3720	3727a	3744	3749
271-280		3751	3765	3785	3799a	3809a	3826	3837	3838	3842a	3861
281-290		3868	3869	3915	3923	3951a	3954a	3956a	4361a	4363	4365a
291-300		4368a	4373	4401	4405a	4406	4410	4413	4417	4508a	4516
1042 301-310	TE	4519	4521	4529	4532	4542	4543	4545a	4547	4561	4564
311-320		4565	4567	4578	4754	4796	4799	4811	4812a	4849	4853
321-330		4882	4897	4903	4916	4946	4962	5006	5027	5033a	5068
331-340		5078	5079	5096a	5129a	5150	5157	5165	5200	5217	5221
341-350		5248	5256a	5259	5265	5271a	5277	5290	5293	5312a	5314
351-360		5319	5338	5343	5358	5360a	5369	5380a	5388	5408	5415
361-370		5420	5424	5425	5446a	5454a	5458	5459	5465	5470a	5477a
371-380		5478a	5483	5490a	5508	5527	5529	5534	5540	5560a	5561
381-390		5572	5583	5594	5597	5610	5620	5627a	5653	5656	5689
391-400		5690	5693	5694a	5703	5705	5710a	5717	5731	5753	5769
1042 401-410	TE	5777	5806	5826	5840	5874	5886	5890	5892	5893	5897
411-420		5907	5917	5919	5921	5923	5925	5926	5933	5945	5948
421-430		5950	5951	5964	5965	5967	5973	5978a	5982	6003	6004
431-440		6007	6019a	6020a	6036a	6040	6043	6044a	6053a	6057	6060a
441-450		6076	6105	6110	6115	6117a	6124a	6137	6139a	6151	6160
451-460		6164	6173	6176	6179	6209a	6222a	6223	6237a	6270a	6274a
461-470		6346	6365	6382	6385	6388a	6399	6423	6432	6433	6644a
471-480		6667	6675a	6683	6695	6696a	6719	6722	6732	6744	6747
481-490		6749	6752	6758a	6761	6763	6768a	6769a	6771c	6775a	6782
491-500		6798	6802a	6822	6832	6834	6838	6858	6868	6892	6910

		1	2	3	4	5	6	7	8	9	0
1042 501-510	TE	6916a	6919a	6921	6944	6952	6955	6960	6962	6971a	7006
511-520		7013	7014	7028	7030a	7043	7054a	7057a	7074	7079	7097
521-530		7123	7133	7134a	7157a	7161a	7164a	7168	7170	7171	7175
531-540		7176a	7185	7229	7231	7234	7250	7305	7306	7317	7320a
541-550		7322a	7324	?	?	7341	7345a	7346	7347	7351	7352
551-560		7356	7367	7373	7376	7386	7395	7399	7404	7410	7411
561-570		7416	7417	7436	7438a	7441	7496	7509	7510	7512	7513a
571-580		7517	7540a	7543	7545a	7546	7551	7552	7555	7562	7571a
581-590		7573	7624	7628	7636	7641	7657	7669a	7695	7709a	7728a
591-600		7743a	7757	7768a	7775	7779	7782	7787	8002a	8003	8014
1042 601-610	TE	8021	8026	8027	8028a	8032a	8034a	3296	4518	5073	6096
611-620		6714	036	1151	2819	2864	4117b	4750	5698	6725	7729
621-623		7730	7732	6190							

Notes:

[a] Locomotives marked 'a' were thought to be still surviving in the USSR in 1984, but no official information about their renumbering has yet been found. All other renumberings are known either from official lists or from observations of locomotives.

[b] The number TE-4117 appears in the list of locos renumbered at Radviliskis depot. 52 4117 was never built, and the correct indentity of 1042 616 remains unknown.

[c] TE-456 and TE-6771 appear in the list of locos renumbered at Radviliskis depot, although two locos with these numbers were sold to Poland circa 1963. It is not known which were the correct 52 456 and 52 6771. Curiously, the renumbering list for the Baltic Railway omits TE-120 which was observed Radviliskis depot in 1992.

[d] New numbers were applied incorrectly to a few locomotives, for example TE-6919 and TE-7057.

[e] Many locos never carried the new numbers.

Appendix I. Principal Dimensions

Maximum speed, forward and reverse	80	km/h
Cylinder diameter	600	mm
Piston stroke	660	mm
Boiler pressure	16	kg/cm^2
Grate area	3.9	m^2
Heating surface of firebox	15.9	m^2
Heating surface of boiler: large tubes	71.3	m^2
Heating surface of boiler: small tubes	90.4	m^2
Total evaporative heating surface	177.6	m^2
Superheater area	63.7	m^2
Water capacity of boiler	7.75	m^3
Driving wheel diameter, with 75 mm tyres	1400	mm
Wheelbase of loco	9200	mm
Axle loading, approx.	15	t

	with bar frames	with plate frames
Weight of loco, empty	76.4 t	76.0 t
Weight of loco in working order	84.4 t	84.0 t
Adhesive weight	75.6 t	75.1 t

Tender	**K 2' 2' T 26**	**K 4 T 30**	**K 2' 2' T 32**
Coal capacity	8 t	8 t	10 t
Water capacity, approx.	26 m^3	30 m^3	32 m^3
Weight, empty	25 t	22.5 t	18.1 t
Weight, laden	59 t	60.5 t	60.1 t
Wheel diameter, with 60 mm tyres	970 mm	970 mm	930 mm
Total wheelbase, loco and tender	18 890 mm	18 230 mm	19 000 mm
Length over buffers, loco and tender	22 940 mm	22 840 mm	22 975 mm

Appendix J. Bulgarian class 15

(Based on material from Stephen Stoichkov, with additional thanks to Assen Stoyanov MD.)

	1	2	3	4	5	6	7	8	9	0
15.01-10	52 508	694	697	699	700	733	1702	1703	2097	2280
11-20	2341	2429	2512	2672	2673	2674	2728	2729	2800	2801
21-30	2803	3236	3237	3240	3241	3242	3243	3774	3775	3777
31-40	3805	3806	3807	3808	3844	3845	3846	3847	3849	3850
41-50	3851	3852	3854	3855	4046	4047	4048	4049	4976	4977
51-60	4978	4980	4981	4983	5961	6254	6342	6349	6425	6426
61-70	6745	6801	7660	7661	7662	7663	7664	7665	7666	7667
71-80	7670	7671	7672	7674	7675	7676	7677	7678	7679	7691
81-90	7692	7694	7699	7700	7706	156	5434	5487	183	6631
91-100	151	4798	5375	4822	1393	247	172	173	4870	2152
15.101-110	52 248	3247	5414	6694	5905	7522d	7487	1114d	7388	6367d
111-120	7484	6672d	6650	7457d	6370	2495	4829	7438d	7519	7527
121-130	1547	4843	1596	6361	7460d	760	3395	3635	1225	1468
131-140	5658	3230	6411	2009	5228	655c	3225d	1229	1301	1695
141-150	2184	1321	3737	3780	3791	3830	3884	5028	5176	56--
151-160	56--	3263	6392	6950	3181b	3178	8016	722	866	1645c
161-170	5074	654	?	683	7440c	738	757	5276	3112	?
171-180	720	7032	2547c	6051	5112	2174	7655	5084	5691	8005
181-190	1341	663	4942	7048	1708	758	2793b	1242	3411	4549
191-200	6283c	4950	744	7338	7353	6854	2182	7363	7368	7539
15.201-210	52 6325	4937	735	8031	4527	3355b	7028c	4520	4566	5131b
211-220	158ÜK	7296	740	6376	5915	6953	3957b	2534	3515	2438
221-230	6791	2145	7514	2833	679	5622	2392	1338	4405b	5301
231-240	7495	7005	2840	3415	6287	6831	?	1518	3582	5613
241-250	3835	a	3352b	4526	1527	3286c	2399	2368b	2732	649
251-260	5579	5151	2223c	7312	2808	6920c	5725	5363	7742	5491
261-270	5525	281	677	1318	6089	7199	2310	3175	4920	6239
271-275	3228c	7523b	6377c	4793	3375					

Also on Soviet list of locos sold to BDZ:
(1960) 52 854, 3439, 3462, 3671, 4938, 5668, 5697, 6129, 6292, 6857, 7379, 7721.
(1964) 52 764, 2810, 3797, 3872, 5170, 5208, 5255, 5342, 5385, 5990, 6201, 7656, 9103.

Notes:
a = ex TE-764? Frames ex 52 2373. Boiler from 52 764.
b = not on Soviet list of locos sold to BDZ. (15.206: Examination of loco shows frames and boiler KM 16481; 15.210: D. Dejanow gives TE-5084)
c = Doubtful. Not on Soviet list of locos sold to BDZ, and conflicting with evidence from other countries: 15.160=52 1708?; D. Dejanow gives 15.207=TE-3797 and 15.256=TE-5342.
d = Doubtful. Conflicts with information from other countries.

Works plate on boiler of 52 4520, later Bulgarian State Railways 15.208. S. Stoichkov.

Preserved locomotives

Column 1: Original identity
Column 2: Current identity
Column 3 & 4: Country and location
Column 5: Owner / other additional details

52 021	TE 021	Lat	Jelgava	Latvian Railway Museum
52 036	TE 036	Lat	Jelgava	Latvian Railway Museum
52 074	TE 074	Belarus		Byelorussian Railway Museum
52 100	52 100	Ö	Straßhof	Railway Museum
52 196	150.1105	Rum	Sibiu	
52 221	52 221	CH	St. Sulpice	
52 335	TE 335	Belarus		Byelorussian Railway Museum
52 360	52 1360	D	Vienenburg	Railway Museum
52 422	TE 5694	Rus	Zelenogorsk	(Stored.) October Railway Central Museum collection
52 460	33-240	Ö	Straßhof	Railway Museum
52 464	520,075	H	Hatvan	
52 622	TE 622	Rus	Shcherbinka (Moscow), Intertrack	
52 662	52 1662	D	Hermeskeil	Dampflok-Museum Hermeskeil
52 777	TE 777	Rus	Kaliningrad	
52 855	52 855	Ö	Floridsdorf	
52 1114	TE 1114	Belarus		Byelorussian Railway Museum
52 1198	52 1198	Ö	Linz	ÖGEG
52 1227	52 1227	Ö	Wien Ost	Brenner & Brenner
52 1423	52 1423	D	Hermeskeil	Dampflok-Museum Hermeskeil
52 1469	TE 1469	Rus	Roslavl	
52 1481	Ty2-1401	Pol	Koscierzyna	
52 2093	52 2093	D	Hermeskeil	Dampflok-Museum Hermeskeil
52 2118	Ty2-2118	Pol	Pyskowice PMP	(Subject to confirmation)
52 2195	52 2195	D	Nördlingen	Bayerisches Eisenbahnmuseum
52 2202	52 2202	D	Prora	Railway Museum
52 2277	Ty2-1387	Pol	Chojnice	
52 2377	33-339	Sln	Jesenice	Slovenian Railway Museum
52 2436	52 2436	Japan		
52 2505	TE 2505	Rus	Babaevo depot.	October Railway Central Museum collection
52 2596	TE 2596	Belarus		Byelorussian Railway Museum.
52 2751	52 2751	D	Marl	
52 2770	2770	N	Hamar	Railway Museum
52 2803	15.21	Bu		
52 2817	Ty2-953	Pol	Chabówka	
52 3109	152 3109	D	Sinsheim	Auto + Technik Museum
52 3162	52 3162	Rus		(Subject to confirmation)
52 3234	33-090	Mac	Titov Veles	
52 3240	15.24	Bu		
52 3314	52 3314	Belg	Haine St Pierre	Chemin de Fer à Vapeur des 3 Vallées
52 3316	52 3316	Ö	Linz	ÖGEG
52 3368	TE 3368	Est	Valga	
52 3397	TE 3397	Rus	Zelenogorsk	(Stored). October Railway Central Museum collection
52 3417	33-253	Sln	Jesenice	Slovenian Railway Museum
52 3504	5621	Lux		G.A.R.
52 3517	52.3517	Ö	Ampflwang	ÖGEG
52 3535	520,030	H	Fertöboz	
52 3548	52 3548	D	Nördlingen	Bayerisches Eisenbahnmuseum
52 3554	26.101	Belg	Brussels	PFT-TSP
52 3644	TE 3644			
52 3816	52.3816	Ö	Lienz	Eisenbahnfreunde Lienz
52 3879	52 3879	NL	Rotterdam	Stoom Stichting Nederland
52 3915	TE 3915	D	Olsberg-Bigge	Gasthaus "Brauerei"

52 4544	52 4544	D	Kassel	Hessencourrier
52 4552	52.4552	Ö	Summerau	ÖGEG
52 4751	TE 8026	Belarus		Byelorussian Railway Museum
52 4752	33-107	It	Trieste	Railway Museum
52 4862	56523	TR	Camlik	Railway Museum
52 4867	52 4867	D	Frankfurt / M	Historische Eisenbahn Frankfurt
52 4900	52 9900	D	Halle-P	Verkehrsmuseum Dresden
52 4924	52 4924	D	Chemnitz	Sächsisches Eisenbahnmuseum (in former depot Hilbersdorf)
52 4936	33-110	Sln	Nova Gorica	Slovenian Railway Museum
52 4966	52 4966	D	Berlin	Museum für Verkehr und Technik
52 4984	52.4984	Ö	Knittelfeld	ÖBB
52 4995	Ty2-1055	Pol	Malaszewicze	
52 5156	520,034"	H	Komárom	
52 5200	TE 5200	Uzb	Tashkent	Railway Museum
52 5415	TE 5415	Rus	Tikhoretsk	
52 5422		Ö	Wien Nord	ÖBB
52 5448	52 5448	D	Leipzig	Eisenbahnmuseum Bayerischer Bahnhof
52 5679	52 5679	D	Falkenberg	Plinthed at depot.
52 5680	Ty2-5680	Pol	Jaworzno	
	TE 5694			See 52 422
52 5804	52 5804	D	Neuenmarkt-Wirsberg.	Deutsches Dampflokomotiv Museum (DDM)
52 5806	TE 5806	Belarus		Byelorussian Railway Museum
52 5865	5865	GB	Bressingham	Bressingham Steam Museum
52 5915	15.215	Bu		
52 5925	TE 5925	Rus	Roslavl	
52 5933	TE 5933	CH		Verein Historische Lokomotiven
52 6115	TE 6115	Ukr	Shchors	Leva
52 6191	Ty2-1035	Pol	Jaworzyna	
52 6340	Ty2-702	Pol	Zdunska Wola Karsznice	
52 6365	TE 6365	Belarus	Lida	Intertrack
52 6644	TE 6644	Rus	Lebyazhe (near St.Petersburg). Stored.	Central Museum of Rly Transport
52 6666	52 6666	D	Schöneweide	DB AG
52 6721	52 6721	D	Hermeskeil	Dampflok-Museum Hermeskeil
52 6744	TE 6744	Rus	Shcherbinka (Moscow), Intertrack	
52 6769	TE 6769	Rus	St Petersburg	
52 6832	TE 6832	Belarus		Byelorussian Railway Museum
52 7046	52.7046	Ö	Selzthal	
52 7047	555.3221	Cz	Nymburk	
52 7102	52 7102	Ö	Wien	
52 7173	Ty2-7173	GB	Nene Valley Rly.	
52 7175	TE 7175	Pol	Malbork	
52 7351	TE 7351	Belarus	Lida	Intertrack
52 7409	52 7409	D	Würzburg	
52 7428	56547	TR	Usak	
52 7434	56553	TR	Denizli	
52 7447	555.3008	SK	Kuty	
52 7468	150.1123	Rum	Dej	
52 7565	555.3254	SK	Bratislava Vychod	
52 7593	52.7593	Ö	Straßhof	
52 7594	52.7594	Ö	Straßhof	Railway Museum
52 7596	52 7596	D		Eisenbahnfreunde Zollernbahn
52 7612	52.7612	Ö	Wiener Neustadt	Brenner + Brenner
52 7620	33-502	Ö	Straßhof	Railway Museum
52 8001		D		Förderverein Historische Kleinbahn, Kalbe (Milde)
52 8004		D	Prora	Railway Museum
52 8006		D	Hermeskeil	Dampflok-Museum Hermeskeil
52 8007		D		
52 8008		D	Jüterbog	
52 8009		D	Jüterbog	
52 8010		NL	Beekbergen	VSM

52 8012		D	Aalen	Royal Erlebnisgastronomie GmbH
52 8013		D	Jüterbog	
52 8014		D		
52 8015		D		
52 8017		D	Berlin	Eurovapor
52 8019		D	Tuttlingen	for planned Railway Museum
52 8020		D	Tuttlingen	for planned Railway Museum
52 8021		D	Jüterbog	
52 8023		D	Jüterbog	
52 8026		D		Eisenbahnfreunde Oebisfelde
TE 8026	see 52 4751			
52 8028		D		
52 8029		D	Röbel (Müritz)	Eisenbahnverein "Hei Na Ganzlin" e.V.
52 8030		D	Jüterbog	
52 8034		D	Simbach / Inn	Plinthed.
52 8035		D	Jüterbog	
52 8036		D	Jüterbog	
52 8037		D		
52 8038		D		
52 8039		D		
52 8041		D		Eisenbahnfreunde Wittenberg
52 8042		D	Jüterbog	
52 8043		D		
52 8044		D	Jüterbog	
52 8047		D		
52 8051		D	Tuttlingen	for planned Railway Museum
52 8053		NL	Beekbergen	VSM
52 8055		D		Eisenbahnfreunde Zollernbahn
52 8056		D	Bautzen	Plinthed
52 8057		D	Tuttlingen	for planned Railway Museum
52 8058		D		
52 8060		D	Walburg	Private purchaser
52 8062		D		Eisenbahnfreunde Wanfried
52 8063		D	Falkenberg	
52 8064		D		Eisenbahnfreunde Wanfried
52 8068				
52 8069		D	Tuttlingen	for planned Railway Museum
52 8072		D	Jüterbog	
52 8075		D		
52 8077		D		Dampfbahn Kochertal
52 8079		D		Schwaben Dampf
52 8080		D	Bautzen	
52 8082				
52 8083		D	Prora	Railway Museum
52 8085		D	Jüterbog	
52 8086		D	Dahlhausen / Wupper	
52 8087		D		
52 8089		D	Jüterbog	
52 8090		D	Hermeskeil	Dampflok-Museum Hermeskeil
52 8091		D		Brenzbahn Museumsverein
52 8092		D	Jüterbog	
52 8095		D	Dieringhausen	Railway Museum
52 8097		D		
52 8098*		D	Nördlingen	Bayerisches Eisenbahnmuseum
52 8100		D	Jüterbog	
52 8102		D	Jüterbog	
52 8104		D	Jüterbog	
52 8106		D	Treysa	Eisenbahnfreunde Schwalm-Knüll
52 8109		D		Thüringer Eisenbahnverein
52 8111		D	Tuttlingen	for planned Railway Museum

TE-5200 at Tashkent railway museum, Uzbekistan, in 1989. Just three years earlier it had been at Kovel depot in the Ukraine. Photo Vadim Goncharov.

52 8113	D	Hermeskeil	Dampflok-Museum Hermeskeil
52 8115	D	Cottbus	Lausitzer Dampflokclub
52 8116	D	Dieringhausen	Railway Museum
52 8117			
52 8118	D		
52 8120	D	Hermeskeil	Dampflok-Museum Hermeskeil
52 8121	D		Eisenbahnfreunde Betzdorf
52 8122	D	Jüterbog	
52 8123	D	Hermeskeil	Dampflok-Museum Hermeskeil
52 8125	D	Tuttlingen	for planned Railway Museum
52 8126	D	Jüterbog	
52 8129	D		
52 8130	D	Tuttlingen	for planned Railway Museum
52 8131	D	Salzwedel	Private purchaser
52 8132	D	Jüterbog	
52 8133	D	Jüterbog	
52 8134	D		
52 8135	D	Diepensee	
52 8137	D		
52 8138	D	Tuttlingen	for planned Railway Museum
52 8139	NL	Beekbergen	VSM
52 8141	D		Property of Landratsamtes Zittau
52 8142	D	Cottbus	
52 8145	D		
52 8147	D	Quedlinburg	Plinthed in front of station
52 8148	D		Pfalzbahn GmbH
52 8149	D	Chemnitz Hilbersdorf	
52 8150	D		
52 8152	D		
52 8154	D	Leipzig	Eisenbahnmuseum Bayerischer Bahnhof
52 8155	D	Jüterbog	
52 8157	D	Jüterbog	
52 8160	NL	Beekbergen	VSM
52 8161	D		
52 8163			
52 8165	D	Hamburg-Farmsen	Plinthed
52 8168	D	Nördlingen	Bayerisches Eisenbahnmuseum
52 8169	D	Tuttlingen	for planned Railway Museum

52 8170		D	Jüterbog	
52 8173		D		
52 8174		D		
52 8175		D	Jüterbog	
52 8176		D	Tuttlingen	for planned Railway Museum
52 8177		D	Schöneweide	Dampflokfreunde Berlin
52 8180		D		
52 8183		D	Schwarzenberg	Verein Sächsischer Eisenbahnfreunde
52 8184		D	Staßfurt	
52 8186		Ö	Linz	ÖGEG
52 8187		D	Jüterbog	
52 8189		D	Salzwedel	Private purchaser
52 8190		D	Prora	Railway Museum
52 8191		D	Tuttlingen	for planned Railway Museum
52 8194		D	Jüterbog	
52 8195		D	Nürnberg	Fränkische Museums-Eisenbahn
52 8196		D		
52 8197		D	Hermeskeil	Dampflok-Museum Hermeskeil
52 8198		D	Tuttlingen	for planned Railway Museum
52 8199		D	Schwarzenberg	Verein Sächsischer Eisenbahnfreunde
52 8200		Belg		Chemin de Fer à Vapeur des 3 Vallées
30-017	33-032	Ö	Straßhof	Railway Museum
30-022	33-037	Sln	Ljubljana	Slovenian Railway Museum
56504	56504	TR	Ankara	
Ty42-1	Ty42-1	Pol	Jaworzyna	
Ty42-9	Ty42-9	Pol	Zdunska Wola Karsznice	
Ty42-105	Ty42-105	Pol	Koscierzyna	
Ty42-107	Ty42-107	Pol	Rybnik	

Other preserved locomotives (with class 52 numbers unknown or requiring confirmation):

52 ???	15.131	Bu	
52 1956 ?	Ty2-7	Pol	Lazy
52 005 ?	Ty2-16	Pol	Stalowa Wola Rozwadow
52 ???	Ty2-29	Pol	Sucha Beskidzka (Schichau 4002)
52 162 ?	Ty2-38	Pol	Jaworzyna (carries number Ty2-88)
52 200 ?	Ty2-50	Pol	Chabówka
52 2112 ?	Ty2-81	Pol	Jaworzyna
52 2175 ?	Ty2-233	Pol	Jaworzyna
52 2520 ?	Ty2-305	Pol	Jaworzyna
52 5605 ?	Ty2-540	Pol	Legnica depot
52 1346 ?	Ty2-911	Pol	Chabówka
52 2794 ?	Ty2-949	Pol	Jaworzyna
52 ???	Ty2-993	F	Paris
52 2733 ?	Ty2-1184	Pol	Chabówka
52 851?	Ty2-1398	Pol	Choszczno
52 163 ?	Ty2-1407	Pol	Elk

Abbreviations

Belg	Belgium	Lat	Latvia	TR	Turkey
Belarus	Belarus	Lux	Luxembourg	Ukr	Ukraine
Bu	Bulgaria	Mac	Macedonia	Uzb	Uzbekistan
CH	Switzerland	N	Norway	ÖGEG	Österreichische Gesellschaft für Eisenbahngeschichte
Cz	Czech Republic	NL	Netherlands		
D	Germany	Ö	Austria	PFT-TSP	Patrimoine Ferroviaire Touristique – Toeristisch Spoor Patrimonium
Est	Estonia	Pol	Poland		
F	France	Rum	Rumania		
GB	Britain	Rus	Russian Federation	*	52 8098 received frames from 52 8101 in 1985.
H	Hungary	SK	Slovakia		
It	Italy	Sln	Slovenia		

Principal Sources of Information

Austria: ÖBB / GKB: Roland Beier, Erich Hoch.

Bulgaria:
(1) D. Dejanow, "Die Lokomotiven der Bulgarischen Staatsbahnen", Vienna, Verlag Slezak, 1990.
(2) D. Dejanow, "Die Lokomotiven BR52 der BDZ", Vienna, Eisenbahn magazine, issue 2/92.
(3) D. Dejanow, S. Stoitschkow, D. Fakes, P. Spencer, R. G. Farr: additional unpublished details.

Germany:
(1) DR locomotive withdrawal dates: Volkmar Kubitzki and Andreas Stange.
(2) Details about classes 52^{80} and 52^{90}: Volkmar Kubitzki.
(3) Locomotive allocations in 1970: from a list published in 1970 by Eisenbahn-Kurier e.V.
(4) M. Weisbrod and H. J. Obermayer: Die Baureihe 52. Eisenbahn Journal Sonderausgabe II/96.

Hungary: G. Villanyi (Budapest) via Roland Beier.

Jugoslavia: C. J. Halliwell, "The Locomotives of Jugoslavia", Frank Stenvalls Förlag, 1973.

Poland:
(1) Observations of chassis works-numbers of more than 160 locomotives.
(2) List of boiler numbers of Ty2s which were obtained from the Soviet Union.
(3) Plan for the renumbering of former German locos, circa 1947.
In comparison with (1) above it is apparent that the list shows more-or-less correctly which locos were in Poland after the war. However the correct Polish Ty2 number for each class 52 loco is in many cases questionable.
(4) List of Soviet class TE locos sold to Poland, with date and location of handover: A. Vassiliev.
(5) Locomotive and boiler log-books.

Rumania: Serban Lacriteanu, Brian Garvin, David Fakes, Roland Beier.

USSR:
(1) Card files of the Ministry of Ways of Communications of the USSR, Moscow.
(2) Locomotive records, St.Petersburg railway museum.
(3) "Manual of scale-drawings for locos type 1-5-0 series TE", Volumes 1 - 3, Moscow, Transzheldorizdat, 1956 - 1957.
(4) Locomotive log-books.
(5) M.D.Samoylov, "Construction, operation and repair of class 52 Locomotives", Vilnius, TIESA, 1949.
(6) V.S.Sharonin, "Transportation and modification of captured locomotives", Moscow, Transzheldorizdat, 1945.
(7) "Railwayman of Byelorussia", Gomel, 1946.
(8) "Soviet Railwayman", Smolensk, 1958.
(9) Central state archive of the National Economy, Moscow.
(10) Boiler records from the Sverdlovsk district.
(11) Offical list of class TE locomotives on the Pribaltic Railway 1984, with new (1042) and old (TE) numbers: received via Toms Altbergs.
(12) Official list of locomotives on the Byelorussian railway in 1958: received from Minsk via Sheridan Mazurek.
(13) Observations of locomotives in the USSR 1953 - 1993: A. Vassiliev.
(14) Observations of locomotives in the former USSR, 1992-1994: S. Mazurek, D. Fakes, M. Murray, B. Garvin, P. Spencer, P. Slaughter, and others.
(15) Observed works numbers of former Soviet locos in Poland and Bulgaria, and information from their official records.

Books:
(1) H. Griebl and H. Wenzel, "Geschichte der deutschen Kriegslokomotiven", Vienna, Verlag Slezak, 1971.
(2) A. B. Gottwaldt, "Deutsche Kriegslokomotiven 1939-1945", Stuttgart, Franckh'sche Verlagshandlung, 1983.
(3) "World War II German Military Studies", Vol 17. Part VII, The Eastern Theatre, p21, New York, Garland Publishing Inc, 1979.
(4) "German Austerity 2-10-0" (Loco Profile Nr 18), Brian Reed, published in Windsor by Profile Publications Ltd, 1971.

Journals:
(1) Jnl. Inst. Locomotive Engineers, Vol XXXVII, London, 1947: pp 278-323, Paper Nr 467: "A General Survey of the German Locomotive Industry during the war years 1939-45"
(2) Ibid, pp 498-523, Paper Nr.471: "Recent German Locomotive Practice".
(3) Various issues of the following magazines reporting locomotives still in use: World Steam; Continental Railway Journal; Railway Scene; Bulletin of the Locomotive Club of Great Britain.
(4) LOK Report. Various issues giving DR withdrawal dates.

Translation: Roland Beier, Anna and Jack Hodgkinson.

German Words

Aluminiumwerk	Aluminium works
Anlieferung	Delivery
Ausbesserungswerk	Main repair workshops
Beschädigt	Damaged
Bestellt bei	Ordered from
Betonwerk	Concrete works
Brauerei	Brewery
Buntmetallindustrie	Non-ferrous metal industry
Chemische Industrie	Chemical industry
Elektrizitätswerk	Power station
Frankreich	France
Gbf	Freight depot
Gebaut von	Built by
Gebiet	Area
Heizlok	Heating locomotive
Jugoslawien	Jugoslavia
Kessel	Boiler
Kesselexplosion	Boiler explosion
Kohlenindustrie	Coal mining industry
Nach	According to/After
Nach einer anderen Quelle	According to another source
Ölindustrie	Oil industry
Ohne Kessel	Without boiler
oder	or
Metallwerk	Metalworks
Nicht abgenommen	Not taken into service
Nicht fertiggestellt	Not completed
Papierfabrik	Paper mill
Rahmen	Frames (chassis)
Reparaturwerk	Repair workshops
Schwefelkombinat	Sulphur combine
Seehafen	Seaport
Siehe	See
Ungarn	Hungary
Wagen Reparatur Werk	Wagon repair works
Waldindustrie	Forestry industry
wurde als xxx abgeliefert	was delivered as xxx
zerlegt	cut-up
Zementindustrie	Cement industry
Zuckerfabrik	Sugar factory

We publish and distribute books about railways in many countries. You are welcome to visit our shop at Föreningsgatan 12 in central Malmö. For a booklist, just send your name and address together with an international reply coupon to

Frank Stenvalls Förlag, Box 17111, S-20010 Malmö, Sweden.

Karte der westlichen Sowjetunion.

¶ Depots der Reihe 52.
Kursiv Reserveabstellplätze.

Das Bahnnetz entspricht dem Stand von 1988.

Map of the western Soviet Union.

¶ *Known depots of the Class 52.*
Italics *Locomotive reserves.*

The map shows the railway network of 1988.